COUNTING

Also by Deborah Stone

Policy Paradox: The Art of Political Decision Making

The Samaritan's Dilemma

The Disabled State

The Limits of Professional Power

Counting

How We Use Numbers to Decide What Matters

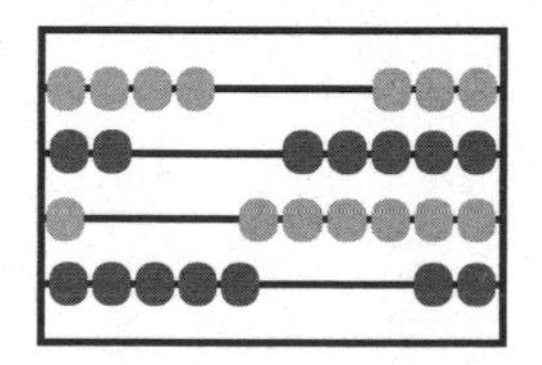

DEBORAH STONE

Liveright Publishing Corporation
A Division of W. W. Norton & Company
Independent Publishers Since 1923

Printed in the United States of America
First Edition

For information about special discounts for bulk purchases, please contact W. W. Norton Special Sales at specialsales@wwnorton.com or 800-233-4830

Manufacturing by Lake Book Manufacturing
Book design by Dana Sloan
Production manager: Anna Oler

Library of Congress Cataloging-in-Publication Data

Names: Stone, Deborah, author.
Title: Counting : how we use numbers to decide what matters / Deborah Stone.
Description: New York, NY : Liveright Publishing Corporation, [2020] |
Includes bibliographical references and index.
Identifiers: LCCN 2020014990 | ISBN 9781631495922 (hardcover) |
ISBN 9781631495939 (epub)
Subjects: LCSH: Statistics—Social aspects. | Counting—Social aspects. |
Measurement—Social aspects. | Evaluation—Social aspects.
Classification: LCC HA29 .S8223 2020 | DDC 001.4/22—dc23
LC record available at https://lccn.loc.gov/2020014990

Liveright Publishing Corporation, 500 Fifth Avenue, New York, N.Y. 10110
www.wwnorton.com

W. W. Norton & Company Ltd., 15 Carlisle Street, London W1D 3BS

1 2 3 4 5 6 7 8 9 0

To count (verb): *to tally, to add up, to total, to recite numerals in ascending order*

To count (verb): *to matter, to be considered, to be included, to have importance*

Contents

Prologue: Of Two Minds

During my sophomore year of college, I had an identity crisis. I loved math and science courses because they reassured me there is some order in the world. Besides, math and science problems had right answers and I was good at finding them. But when daytime coursework passed into late-night dorm sessions, literature, philosophy, and politics posed more urgent and discussable questions: What is the meaning of life, and how can people live together to make the best of it? These questions had no right answers and my grades told me I wasn't very good at dealing with ambiguity. But at age 18, those were the questions I desperately needed to pursue.

I was trapped between "the two cultures," Sir Charles Snow's phrase for the chasm he observed between scientists and literary types. The two camps simply couldn't talk to each other.

They spoke different languages, they thought differently, and they certainly had different ways of pursuing truth. My father had given me Snow's book in high school. At the time, it didn't speak to me. Now I saw that Dad had sensed my struggle long before I did.

Browsing one day in a campus bookstore, I found a ray of hope in an essay called "The Creative Mind" by Jacob Bronowski. Bronowski challenged the two-cultures divide by showing that creativity is the same process for scientists, poets, and painters. They all grope for new understanding by finding "hidden likenesses" that others haven't noticed. Scientists, he wrote, don't make discoveries by "taking enough readings and then squaring and cubing everything in sight." Copernicus couldn't have gotten the idea that the earth revolves around the sun with only a camera and a measuring stick. "His first step was a leap of imagination—to lift himself from the earth, and put himself wildly, speculatively into the sun." From that vantage point, he saw that "the orbits of the planets would look simpler if they were looked at from the sun and not the earth." For Bronowski, Copernicus's creative moment, the poet's metaphor, and the painter's imagery are all of a kind.

I felt liberated by Bronowski's insight. I didn't know how I was going to reconcile the two cultures for myself, but I knew it would require some leaps of imagination. I bolted from my science path and through a circuitous route became a political scientist, despite an ominous note from a professor on one of my papers: "B—This is a credible effort, but you'll never be a political scientist." Political science is an oxymoron if ever there was one. The very name spans the two cultures. I should have known I would remain caught between the two cultures for the rest of my days.

I landed my first teaching job in another oxymoron, a new program called "Policy Science." My colleagues were positively smitten with numbers. There was no policy problem and no personal problem (Should I marry my girlfriend or boyfriend?) for which statistics couldn't find the best answer. Almost everyone else on the faculty was teaching our students how to find the mathematically best answer to policy problems. As the only political scientist, my job was to teach them how to get their neat solutions through the messy political process. I didn't know much about statistics but I sure knew they don't carry much weight with politicians. I knew that stories are more persuasive than numbers. And I knew that politicians, advocates, and activists could use numbers for both good and evil.

On the side of good, there was asbestos. In the 1960s, labor unions and doctors used numbers to document that breathing asbestos fibers causes mesothelioma, a fatal lung cancer. Finding the link between asbestos and cancer was one of many public health triumphs indebted to numbers. On the side of evil, there was the Vietnam War. Robert McNamara, Lyndon Johnson's defense secretary, measured success in Vietnam by counting dead Vietnamese people, the infamous "body counts" of the nightly news. Over the course of my career, I've seen numbers be touted as either Jekyll or Hyde. At one moment, numbers are the only facts we can trust. At another, there are "lies, damned lies, and statistics."

Nowadays, the battle between the two cultures often takes place under the banner of "Numbers Versus Stories." My town, Brookline, Massachusetts, hosts the first marijuana dispensary in the Greater Boston area. As the town was getting ready to license a second one, nine hundred residents signed a petition asking the town to put restrictions on both dispensaries. At a public hearing, neighbors complained of litter, public urination, and "consuming in public." They spoke about crowded sidewalks, traffic congestion, and parking problems, all harmful to nearby businesses. One woman said she felt afraid to walk home at night.

"Stories aren't facts," the president of the first company responded. "We take the concerns of our neighbors very, very seriously and will continue to do so, but facts and data really need to drive this discussion." By "facts" and "data," she meant numbers. She was staking a familiar claim: People are hopelessly subjective and biased. They see the world from their own narrow point of view. Their anecdotes are nothing but fleeting glimpses of their distorted perspective. Numbers, by contrast, are objective.

I'm all in favor of the pot shops. The "stories aren't facts" gambit bothers me, though, because it dismisses citizens who can't express their experience in numbers. Calling for more data makes a great stalling tactic, too. It tells town officials, "Stick with the status quo until we get some numbers."

On the other hand, what's not to love about objectivity? Objective means fair and impartial—the opposite of biased and playing favorites. For scientists, objectivity means that when different people study the same problem or count the same things, they'll get the same answer. They'll get the same answer because, if they're good scientists, they strip themselves of personal biases. In the words of someone who bills himself as a science advocate, "There is a world. It is real. It is home to objects and processes that exist independent of us and our beliefs." I love that world. It's the world where I can spend hours at a pond contemplating the beavers and frogs, the wind gusts and water ripples, the trees, the clouds, and an oncoming storm. None of those "objects and processes" gives two hoots about what I think of them. That world is a physical world. It is home to things like rocks and tornadoes that don't morph according to how we feel about them.

This book is not about that world. This book is about the social world. The social world is filled with human ideas and experiences that most definitely shapeshift in response to our thoughts and feelings. This book is about how we take the measure of those nebulous notions connected to the meaning of life that goaded me in college and still do. How do we count how much freedom and equality we have? How do we decide who is poor or disabled and deserves society's help? How do we measure democracies to know how democratic they are and how we can make them more so? How do we measure pain so doctors can help us cope with it? How do we measure students' knowledge and teachers' teaching ability to find out how the two are connected? How do we count racial and ethnic identity, and why *do* we count it, anyway? How do we measure the size and strength of a nation's economy? How do we find out how many people are unemployed? How do we count violence and crime?

There's no way to take biases out of these questions because they're not objects independent of us and our beliefs. They're ideas and interpretations of our experience. We fashion and interpret these not-things in the workshops of culture, religion, gender, race, and politics.

In my quest to bridge the two cultures, I've come to think that reimagining numbers is the key. Numbers are, so to speak, the tiniest particles of scientific thinking. They seem like hard-core facts: 1 is 1 and 3 are 3 and there are no two ways about it. But 3 what? Three pieces of freedom? Pasting a number on an idea doesn't tell you what you've got in your hand. The number only makes you believe that you caught whatever you were trying to catch by counting.

Numbers spring from leaps of imagination, from seeing likenesses between things that aren't exactly the same. To count how many people are unemployed, you've first got to decide what unemployment means. You've got to find the hidden likeness between a worker who was displaced by his inability to compete with a robot and one who was displaced by her inability to afford childcare. Our current way of counting the unemployed doesn't see this likeness. We fight about how to measure unemployment or anything else by making imaginative leaps and trying to get others to leap along with us. Those imaginative leaps are metaphors, the same metaphors that are, so to speak, the tiniest particles of literature and art.

To see numbers as metaphors doesn't render them useless. Like all tools, numbers are much more useful if we know what they're made of and how they work. Then we can use their underlying metaphors as springboards for our own imaginative leaps. We can't do *without* numbers, so the challenge is how we can do a lot better *with* them.

In the social world, numbers aren't mere figures. They're authority figures. For centuries, people have wondered why readers and theatergoers suspend their critical faculties long enough to enjoy living in a made-up world. I wonder why people (including me) so often trust numbers, seeing as how finagling with numbers is a common way to cheat. I shudder at some of the ways we allow numbers to determine our own and other people's fates. And I marvel at how the very process of counting can change how we behave. Anyone who's dieted or worn a Fitbit knows how that works.

If every number begins with a judgment, and if we allow numbers to determine people's fates, we should hold numbers to the same ethical standards we hold our judges to. We should expect those who count to disclose and justify their judgments. We should make sure people harmed by counting-decisions have opportunities to challenge them. The more we outsource decisions to automated counting systems, the more important it becomes to authorize humans to make exceptions. Above all, we shouldn't use numbers to cover over our deep conflicts and ethical dilemmas. That's what the Founding Fathers did. They reconciled the North and South by counting slaves as three-fifths of a person in the federal census. Some questions can't be answered with a number. To count well, we need humility to know what can't or shouldn't be counted.

Every chapter in this book tells stories about the life of numbers. The stories illustrate how numbers spring from their creators' imaginations, how their creators infuse them with meaning, and how they wield power. Almost every example could fit nicely in a different chapter from the one where I put it. That's because there are many ways to tell a story about anything, and numbers are no exception. If you see other stories about my numbers besides the ones I tell, as an author I couldn't ask for more.

Brené Brown studies vulnerability and shame. She's an academic researcher who collects people's stories as her data. "Stories are just data with a soul," she says. Too often numbers are used as data without a soul. But numbers are made of stories. They *are* stories. This book is about how to put the soul back into numbers.

COUNTING

1

There's No Such Thing as a Raw Number

When I began work on this book, I thought I should consider how children learn to count, so I headed straight for *One Fish, Two Fish*. If anyone could make counting fun, surely Dr. Seuss is the man. He has a way of getting in league with kids against grown-ups and all their seriousness. But I had a surprise coming. Here's how the book begins:

One fish, two fish,
Red fish, blue fish
Black fish, blue fish
Old fish, new fish
This one has a little star,
This one has a little car
Say! What a lot of fish there are.

Say! What happened to my counting lesson? There are no number words after two. We get lots of fish *qualities* but wind up with the vague quantity term "a lot." It's as though Dr. Seuss has already given up counting after the first line, because he's enchanted by something special about each fish. But let's keep going. Maybe he'll eventually get beyond two.

Some are fast.
Some are slow.
Some are high.
And some are low.
Not one of them
is like another.
Don't ask us why.
Go ask your mother.

Well, no help with counting here. *One Fish, Two Fish* is not a learn-to-count book, as I'd remembered. Dr. Seuss meant this book as a message about tolerance, celebrating differences, and having fun together. But for me, *One Fish, Two Fish* is also a profound statement about why it's impossible to count objectively.

Dr. Seuss never asks, "What makes them all fish? What is fish-ness?" That's the unanswered question and the paradox we can't resolve. Dr. Seuss *ignores* differences to count them all as fish at the same time as he *celebrates* differences to count them all as fish. Here is the existential dilemma of counting: How can we possibly count things if not one of them is like another?

And here is where counting meets power. The only way to count is to force things into categories by ignoring their differences. By the time we're adults, we've lost sight of the mental coercion entailed in counting. So let me refresh your memory by taking you back to preschool.

First, we teach you the sequence of number words in your native language—one, two, three. When you can recite those, we start pointing to objects and saying number words as we point. In the earliest counting lessons, adults put objects or pictures in front of kids, implicitly telling them, "These are all the same. They each count as one member of a group." They're all fish.

But counting is more than merely attaching number words to a heap of things. Before you can tally up how many "somethings" there are, you have to sort things and decide which ones belong to the group of things you want to count, unless an adult has already done the sorting for you. It's one thing to put a row of peas in front of a kid and ask her to count them. It's quite another challenge to give her a heap of peas and soybeans and ask her to count the peas.

We start training children very early to distinguish between what counts as important and what doesn't. A classic type of counting worksheet shows several triangles, rectangles, circles, and stars. Each shape comes in different sizes and colors—small stars, big stars, red stars, blue stars, and so forth, rather like *One Fish, Two Fish*. The point of this exercise is to teach kids to classify before they count. A teacher, or perhaps the worksheet, asks, "How many triangles are there? How many stars?" A child can't do this task unless a grown-up has already taught her some clear-cut rules: "If it has 3 points, it's a triangle. It doesn't matter what color a shape is or how big or small. If it has 3 points, it's a triangle." Along with learning the shape rules, the kid is learning to ignore color and size. The kid is also learning to stereotype. All things with 3 points get treated as triangles, even though some are red and some are blue and some are not very big at all.

Counting, then, entails two mental moves: first classifying, then tallying. In the first phase, counting is a way of making metaphors, because we start by finding similarities among things that are different. Numbers—those things people revere because they're so precise and objective? We construct them by making our own decisions about how to separate things into groups. In the split second before we decide, the thing could go either way; it could be a *this* or it could be a *that*. Numbers are a magic wand that resolves ambiguity into one-ness.

Learning to count is much like learning to talk. The first step in learning to talk is learning to name things. When kids learn the word for "nose," they learn to point to "my nose," "Daddy's nose," "Mommy's nose," "Fido's nose," and to think of all these different things as belonging to the same category, never mind that they look, feel, and *are* different. That wet thing at the tip of Fido's snout is like this little knob on my face and that pointy thing on Mommy's face. They all have the same name so they're all the same. Language is all metaphor.

The connection between naming and counting leapt to mind when I overheard a snippet of conversation in Venice. I was on a vaporetto gliding down the Grand Canal toward the giant domed church called La Salute. In front of me sat a French couple with their little boy, about two, I guessed. As we approached La Salute, he pointed to it and exclaimed, "La

Tour Eiffel!" Perhaps to him the words Eiffel Tower meant something like "big," or "the tallest thing around." How else to explain his thinking except that some concept of size was embedded in his understanding of the name? He was learning to classify objects according to one feature—in this case size—and name them accordingly. And if I'm right, then the way he ignored shape, color, and everything but size to name the building was no different from how a child follows a counting rule such as "If it has 3 points, it's a triangle." We use the same principle to name and to count. Naming is another way of "counting as."

Counting and naming both require us to find similarities between things that are different. Children's learn-to-count books subtly teach about classification along with tallying. In the Sesame Street book *1 2 3 Count with Elmo*, Cookie Monster stands next to the numeral 6 with his paw in the cookie jar, smiling at 6 glasses of milk and 6 cookies. The Count points to the numeral 9 while he savors a night sky with 9 lightning bolts and 9 bats. The milk glasses and lightning bolts are exactly the same in size, shape, color, even their orientation on the page, but no two cookies are alike and no two bats are alike. Some cookies have much more enticing frosting, with pink squiggles, chocolate dots, cherry jam, or sprinkles. Others are more plain. The bats are different colors, some are smiling and some have closed mouths, and some of them have teeth. To a kid, those differences are worth considering. But kids are supposed to learn that even though the cookies look different and taste different and some of them are prettier than others, they're all

cookies. Even though some bats look like better playmates than others, for purposes of counting, they're all the same.

Here, at last, is the lesson you didn't get in preschool. When kids learn to count, they're not just learning number words and numerals; they're learning *how adults see things* and the unspoken rules adults use to consider some things alike and some different.

The U.S. Census Bureau counts racial and ethnic identity using the unspoken rule in the Obama cartoon. Until recently, the census mainly assigned people to five categories that were first proposed by a German doctor in 1776: Caucasian, Asian, Pacific Islander, African American, and American Indian. The Census Bureau hired people to go door-to-door counting noses and marking down a race for each person. Census officials, like most people at the time, imagined the categories as biological facts, but their written instructions for the census takers revealed the sheer political power behind the categories. In 1930, a door-to-door canvasser was told: "A person of mixed white and Negro blood should be returned [counted] as a Negro, no matter how small the percentage of Negro blood." This was the infamous "one-drop" rule that Southern states used to consign Negroes to lesser legal status and discourage interracial marriage.

To census officials of the time, Negroes were easier to classify than Mexicans. "Practically all Mexican laborers," the instructions continued, "are of a racial mixture difficult to classify." To help the census takers, the instructions offered a clear rule: "*It has been decided* [by whom, we don't know] that all persons born in Mexico, or having parents born in Mexico, who are definitely not white, Negro, Indian, Chinese, or Japanese, should be returned as Mexican (Mex)." Clear? Put on your census-taker hat and try implementing that definition.

Gradually, the Census Bureau added more race categories and gave people more freedom to classify themselves. Starting in 1960, the bureau mailed questionnaires for residents to fill out. Enumerators visited only households that either didn't have a mailing address or didn't return the questionnaire. In 1980, the bureau added a new question, separate from the race question, asking people whether they consider themselves Hispanic—yes or no. (How and why that happened is another story, which we'll get to in Chapter 5.) Then, in 2000, for the first time the race question gave people the option to check more than one race. Now there were at least 14 races to identify with, plus a "Some other race" box with space for writing in one's preferred race label. No longer was race something to be assigned by an unnamed authority. Nor was race an all-or-nothing proposition where a person would be counted as having one race regardless of mixed ancestry. Or so it seemed.

"Census: White Becoming Minority in the U.S."

New York Times, June 20, 2018

"The U.S. Will Become 'Minority White' in 2045, Census Projects"

Brookings Institution, *The Avenue* blog, March 14, 2018

Headlines like these come about because when the Census Bureau predicts the future, it still assigns people to race categories using the one-drop rule. People who embrace a mixed heritage by checking "White" and any other box in the race question aren't counted as white. When mixed-heritage couples fill out the census form for their children, they usually try to honor both parents' heritages by checking multiple boxes. That means most children of mixed marriages aren't counted as white. Because intermarriage is increasingly common and children of mixed marriages count as nonwhite—presto!—the white population is going down. And that's not all. The only people who count as whites are those who say they're *not* Hispanic and check *only* the "White" box in the race question. They're called "non-Hispanic whites," as if Hispanics can't be real whites. Everyone else, including Hispanics who say they're white, gets counted as nonwhite and labeled as "Minority." Presto! The U.S. will soon be a "minority-white" nation.

The Census Bureau's counting rules do a number on racial and ethnic integration. Intermarriage is a sure sign of social assimilation. By counting children of mixed-race marriages as nonwhite, the bureau segregates them statistically. It's as though census officials refuse to see progress in social harmony.

How we count makes all the difference—and therein sleeps a giant conundrum. We want to believe numbers are objective, yet we know statistics can lie.

It's easy to see why many people believe numbers are objective. When we first meet them in childhood, they're introduced to us as creatures with impeccable credentials. They're precise, accurate, and beyond dispute. In preschool we learn that there's only one way to count. In elementary school, we learn that every arithmetic problem has one right answer.

As we grow up, we gain another reason to put our faith in numbers: we want to be treated fairly. When people make decisions about us, we want them to decide for good reasons. We want them to be objective. If people with power over us consider only measurable factors in their decisions, we feel somewhat protected against their whims, prejudices, and whatever else might be eating them on the day they decide our fate.

Because numbers enjoy an aura of objectivity, it's tempting to resolve political and moral conflicts by translating them into arithmetic problems. Perhaps no area of law is as thorny and contentious as equal treatment. Title IX of the Civil Rights Act concerns equal educational opportunity for women and girls. One section requires colleges and universities to provide equal athletic opportunities to male and female students. In 2010, a case against Quinnipiac University came down to a classification question similar to whether a cookie with plain white frosting counts the same as one with fancy pink squiggles. The question was: Does cheering and tumbling count as a varsity sport? If cheering and tumbling *does* count, then the university had an equal number of varsity teams for men and women and met its obligation to provide "equal opportunity."

A federal judge ruled that cheering and tumbling couldn't be counted as a varsity sport. To achieve parity, the school would have to either add another women's varsity team or eliminate one of its men's teams. There were other issues in the case, including whether the school accurately reported the number of female athletes on its teams. Still, you see the point. Is there an objective, truthful answer to whether cheering is a sport? Or whether a school provides equal opportunity for men and women in sports? Nope.

Most of the big questions we care about, like equal treatment, are hard to wrap our minds around. If we break them down into counting problems, we feel as if we've tamed them and made them manageable. Then again, most of us know better than to trust numbers. As the saying goes, if you torture the numbers long enough, they'll confess to anything. Stuart Rice, who served as president of the American Statistical Association and as a high official in the Census Bureau, once warned: "Statistical method and statistical data are never ends in themselves. They are always accessory to some purpose." You can almost hear Sherlock Holmes tiptoeing into a conference of statisticians to suss out Nefarious Purpose.

Sixty years ago, a little-known magazine writer made it his business to suss out statistical perps in a book whose title you probably know, even if you haven't read it: *How to Lie with Statistics*. Darrell Huff introduced his book as "a sort of primer in ways to use statistics to deceive." He explained: "It may seem altogether too much like a manual for swindlers. Perhaps I can justify it in the manner of the retired burglar whose published reminiscences amounted to a graduate course in how to pick a lock and muffle a footfall: The crooks already know these tricks; honest men must learn them in self defense."

Huff showed how crooks could distort honest statistics with deceptive packaging—things like biased samples, weasel words, misleading graphics, and taking numbers out of context. But the problems with statistics run far deeper than Huff imagined. Statistics aren't born with honest meanings that people later corrupt with false advertising. Every number is born of subjective judgments, points of view, and cultural assumptions. Numbers are filled with bias through and through, because that's what categories do. Categories are ways of seeing and *not* seeing, in the same way a racist sees skin color without seeing a person.

There's no way out. Categorizing cooks the numbers—not in the sense of deliberate fudging, though there's plenty of fudge to go around, but in the sense that someone has to make judgments and interpretations before counting can begin. Numbers are products of our imagination, *fictions* really, no more true than poems or paintings. In this sense of fiction, all statistics are lies.

One of my friends read a draft of this chapter and protested, "I'm a physicist and I don't think all statistics are lies." His comment brought me up short. Physicists, I quickly realized, deal with the physical world. In that world I'm quite trusting of statistics, too. Every time I drive over Boston's Zakim Bridge, I thank the statistics that proclaimed how strong the suspension cables should be to keep me from plunging to my death. I'm not a physical scientist, though. I'm a social scientist. I deal with the social world. Social scientists truck in ideas and emotions, motives and intentions, hopes and fears, cooperation and conflict, love and hate—and a lot more that I've left out. The social world certainly has elements of the physical world. We live in physical bodies, we walk on the earth, and we sense the physical world every minute with our eyes, ears, noses, tongues, and skin. The physical world is crucial to our survival. Yet as fascinating as that world is, the social world fascinates me more. I'm drawn to the mysteries of social relationships and the puzzles of philosophy. I'm trying to find my way through a world of intangibles. In that world, counting only gets us so far.

In Rachel Kushner's short story about a prison, Gordon Hauser has just started a new job prepping inmates for the GED test. Hauser poses a word problem to one of his tutees: "If there are five children and two mothers and one cousin going to the movies, how many tickets do they need? (a) seven, (b) eight, (c) none of the above."

"What movie are they going to?" the woman asks.

The tutor tells her, "That's the wonderful thing about math: it doesn't matter. You can count without knowing the details."

The woman persists. She says she needs more detail in order to "imagine these people." Hauser has already pinned a judgment on her. He thinks she's a tad obtuse and can't grasp abstraction, so he simplifies the question for her by stripping away some of the details: "There are three adults and five children. How many tickets do they need?"

"You didn't say if they let kids in free, so how can I know how many they need? And depending on what kind of people they are, what theater this is—are they ghetto or are they squares like you? Because maybe they let one of the adults, like that cousin, in through the emergency side door, after they pay for two tickets."

In math class, no matter what kind of thingamajigs you're counting, 5 + 3 = 8. Details about the 5 and the 3 don't matter. Outside the classroom, details matter. Lopping off the details of human experience is like amputating an arm or a leg. Before we amputate, we'd better make sure the patient will benefit.

The inmate-narrator in the story doesn't need remedial first-grade arithmetic. She's sick of being on the butt end of condescension. Instead of telling the tutor straight out, though, she outsmarts him with an excellent tutorial on counting. She teaches him not to start measuring a situation until he understands how people think and behave. And by the way, he ought to learn some details about *her* before he jumps to conclusions about her smarts.

Abstraction, leaving out the details, is a powerful thinking tool. Leaving out the details lets us see the unadorned essence of things. Without abstraction, Dr. Seuss's fanciful fish, the red one, the blue one, the one with a car, and the one with a star—they wouldn't all be fish. Without abstraction, we couldn't compare people's economic situations because when it comes down to the details of income and expenses, "not one of them is like another." With abstraction, we lose important information unless someone like Dr. Seuss gives us the information in words.

Think about Amazon's customer product ratings. A row of five stars sits just beneath most product names. At a glance, you can see a product's overall score, from 1 to 5 stars. When I'm browsing, a 1- or 2-star overall rating kills the deal, but if I'm halfway interested in a product with three or more stars, I click to read the words. Often the words are useless. "I bought this coffee maker for my wife. She loves it." More often than not, though, the verbal comments tell me what people like and don't like about the product and its seller. Numbers help me winnow my choices, but words let me decide what matters to *me*.

Abstraction is more than a thinking tool. It can be a powerful political tool, for better and for worse. It can be a way of stereotyping, as the prison inmate knows so well. It can also start people agitating for political change. Abstraction enables people to conceive of human rights that apply to blacks as well as whites, women as well as men, and fetuses as well as infants. Try lumping women and men together to count them both as humans and you'll get pushback, maybe not in *your* kitchen, but in a sharia court you'll feel the crush of opposition. Don't count fetuses as humans and you'll get an angry protest at an abortion clinic.

Amputations hurt people when they're carried out with machetes by warring tribesmen, but they can help people, too, and usually do when they're performed by surgeons. Antidiscrimination laws could be seen as salutary amputations. They eliminate race, gender, and religion as "details" that can be used to decide whether to hire people, sell them homes, or arrest them.

We can't escape abstraction nor should we try, but we should be careful how we use it.

Trying to measure abstractions such as equality or human rights puts us in peril of the double-edged sword, the one with a blade that cuts two ways. Here's a double-edged sword from Afghanistan. Afghanistan's Independent Human Rights Commission aims to stop human rights violations, with a special focus on gender-based violence. Monthly reports from around the country show that the number of complaints has been increasing from year to year. So has the number of complaints that the commission finds to be genuine human rights violations. One way to interpret the bigger numbers is that the human rights situation is getting worse. But the commission leaders see the increase as a big success. They have worked to educate people about their rights and to encourage people to report violations. To the commission, more reports mean that the culture of human rights is getting stronger. Which is it? Do bigger numbers mean failure or success? I go with the commission's interpretation, because you can't change people's actions without first changing their ideas and attitudes.

Here's another double-edged sword. If a doctor keeps you waiting for an appointment, is that a sign of good care or bad? In my informal polls, most people are annoyed by lateness, especially when it interferes with their own busy schedules. They think a late doctor is an inefficient doctor and should be graded accordingly. Other people—I'm one of them—take running late as a good sign. It means the doctor spends as much time as each patient needs. For these people, a late doctor gives better, more dedicated care and should get bonus points. It's hard to say whether lateness means good care or bad.

How about this double-sided slicer: Which store offers more value for the money, the one with lower prices or higher prices? Low prices mean affordability for consumers. A dollar goes further in a store with low prices than it does in one with higher prices. But low prices might also mean low-quality goods and low wages for employees, some of whom also shop at the store. Are low prices a good measure of value? It depends on your point of view. Are you a consumer or an employee? What if you're both?

At the risk of hacking you to pieces, here's one more double-edged sword. Foreign-aid agencies would like to target their aid to democratic countries. For that, they need to measure democracy. They need to know how much of it nations have and whether a nation has advanced or backslid on the path to democracy. How do they find bits of democracy they can observe and count? Nearly everyone in the democracy-assessment business thinks elections are an essential element of democracy, but they don't always agree on what makes some elections better than others. Some experts think that if a country requires its citizens to vote, higher turnout makes elections more representative; thus, mandatory voting enhances democracy. Others think mandatory voting impinges on citizens' freedom, thus diminishing democracy. Is there a right answer? No, it's a thorny question of "counting as." There are good reasons to count both ways.

In the physical world, things carry on apart from our beliefs about them. It might not be easy to count subatomic particles, but at least they exist outside our heads. Much of what we care about in the social world exists only inside our heads. Sure, events such as doctor visits and elections take place outside our heads. We can count those events rather easily, but measuring what they mean is a different kettle of fish.

For understanding human experience, purely subjective measures have a lot to offer. Unless you've led a blissful, pain-free existence, you've been asked by doctors to rate your pain on a scale from 1 (barely noticeable) to 10 (worst imaginable). Most people I know find this exercise difficult and resist it. "I *hate* that question," a friend told me when I mentioned I was writing about pain scales. We were enjoying a pain-free walk in the Arnold Arboretum, but I stopped admiring the trees to ask her, "Why do you hate it?"

"It's so subjective," she answered. I pressed her to explain.

"My tolerance isn't the same as another person's."

I probed some more: "Isn't it helpful to let doctors know how much *you* can tolerate?"

She thought for a minute. "Maybe it's not hurting at all when I go into PT," she started, "but when they touch me or move me, it hurts like hell. So which number is it?"

Pain is the most isolating experience there is. No one else can know how your pain feels, much less feel it. As crude as pain scales are, they help us to communicate the inexpressible, not only to get sympathy but, ideally, to get help.

Ronald Melzack was a graduate student at McGill University in the 1950s when doctors there developed an instrument to measure pain objectively. As you might suspect, the "dolorimeter" was an instrument of torture, albeit mild. Researchers focused an intense light on subjects' skin, noting when the subjects first cried "Ouch" and when they couldn't stand it anymore and pulled away. In one study, researchers asked pregnant women to say when the pain produced by the light on the back of their hand was equal to the pain of a labor contraction. (Now there's an algebra formula I bet you never learned in school.)

Melzack had the good sense to know that "a tiny burn is not like a headache, a toothache, a heart attack or a kick on the shin"—nor like giving birth—and he quickly rejected the reigning idea that pain could be measured on a simple 1-to-10 scale. During a year as a postdoctoral student, his mentor sent him to a pain clinic and told him to "listen to patients suffering from more pain than you can imagine." The patients, Melzack later wrote, had "a rich vocabulary" to describe their pain in vivid detail. Melzack started collecting words and sorting them into categories. One category described sensory experience, such as pulsing, shooting, pinching, gnawing, tugging, burning, and stinging. Another category included emotion words, such as exhausting, sickening, terrifying, cruel, or wretched. Another he called "evaluative": annoying, troublesome, miserable, intense, and unbearable. Then there were miscellaneous words, including tight, squeezing, cold, freezing, nagging, and torturing. I don't understand why he grouped some of the words the way he did, but that's not important. With all these ways of describing pain, Melzack realized, it was no longer tenable to believe that all pain is experienced the same way, varying only in intensity from "barely noticeable" to "off the charts."

Melzack developed what became the widely used McGill Pain Questionnaire, a sort of thesaurus for pain words. It groups words in an ingenious way that allows patients to communicate the quality of their experience much better than they can with a single number. As another plus for his system, words such as burning, stabbing, or cramping can sometimes enable doctors to pinpoint a diagnosis. Patients' answers to the questionnaire can also tell doctors whether a treatment helps reduce specific kinds of pain. Melzack and his colleagues went on to create numerical scales from the words. But unlike the simple 1-to-10 scale, Melzack's scale requires clinicians to let patients go through the entire questionnaire, checking off all the words that apply.

Words are better than numbers for helping patients overcome the loneliness of pain, and they're better for helping doctors figure out what's causing the pain. Yet almost all medical clinics use the 1-to-10 scale. I once asked a physical therapist why. She knew Melzack's scale well and said she preferred it to the number scale. "But," she added without a moment's hesitation, "insurance wants numbers."

As I talked with more people about the 1-to-10 pain scale, I learned how it has become a covert medium of communication. One friend who struggles with cancer pain said matter-of-factly, "They don't want you to be above a 5." I wasn't sure what she meant.

"Do you mean they try to influence your number?"

"No," she said; "above a 5 means your pain isn't under control. They will want to do something about it."

"So you mean before you give a number, you think about what they want to hear?" I asked.

My friend is a social scientist and intellectually curious right down to her own suffering, so it didn't surprise me when she answered by blurting out the title of a famous sociology book: "It's *The Presentation of Self in Everyday Life*." Each time she visits her clinic, she said, she is asked to rate her pain on a scale from 1 to 10. Before she answers, she thinks about how she wants to appear to her medical team and what actions her answer will trigger. About a year after this conversation, she brought me more news from the pain front. "Now they don't even bother asking me the full question. I sit down and they say, 'Number?'" I wish I could convey her cold, brusque tone and her look of disgust.

Since that first conversation, I've thought a lot about "the presentation of self" when people answer the pain question. A patient might think to herself, How wimpy or stoic do I want to appear? Do I want to trigger a dose increase? Or do I want to hold off and save the big guns for later because I know the pain is probably going to get worse? Even though my pain is unbearable, should I give a number somewhere in the middle to leave room for things to get worse? Should I give a lower number than I gave last week to signal to my doctor that she's succeeding, so she won't give up on me?

In medicine, insurance, and drug research, clinicians have to document results with objective evidence. As my physical therapist indicated, subjective pain numbers now pass for objective measures. But patients know better. One of my students spelled it out for me: "When I was in the hospital for pancreatitis caused by a reaction to chemotherapy, I was on a Dilaudid drip. I quickly figured out that when they asked me for my pain score, the number I should give them was how much more opium I wanted, rather than how much pain I was in. Obviously these two are very different, because sometimes even if I was experiencing pain, that was preferable to being zonked and staring at the wall for days on end." Once I heard this story, I continued asking friends about their experiences with the pain scale. Everyone who'd had a problem for which they needed heavy-duty painkillers told me a version of the same story: patients use the numbers to assert control over their clinicians.

Of the many ideas and experiences we try to measure with numbers, pain falls at the toughest, most subjective end of the spectrum. To be sure, pain has a physical component that doctors can measure by testing how nerves conduct signals to pain receptors. As Melzack insisted, though, pain is an experience with many meanings. The 1-to-10 pain scale is far too crude to capture pain in any meaningful way. Patients have learned how to use those crude numbers as a language for telling their stories.

Numbers serve as a language for telling stories even for more concrete situations such as having a job or being unemployed. On the first Friday of each month, the federal government issues its new jobs report. There's one number—say, 201,000—that's supposed to be a tally of how many new jobs were added to the economy during the previous month. To get the jobs number, the Bureau of Labor Statistics surveys about 150,000 businesses and government agencies and asks them how many new people they hired during the previous month. From there, statisticians use some fancy math to estimate "new hires" for all employers in the country. But what does the jobs number actually mean besides what the pundits say it means? ("Job growth was strong" or "weak but respectable.") Never mind the guesswork that goes into projecting from a small sample to the entire economy. We have a fish problem:

Some jobs are fast and some are slow.
Some jobs pay high and some pay low.
Not one of them is like another.

Well, okay, some jobs *are* like others in ways that matter, but when it comes to personal well-being, the differences are more important than the similarities. Some jobs provide such mind-boggling salaries and stock options that it's impossible to imagine how anyone could possibly spend it all. Facebook founder Mark Zuckerberg is said to have lost $2.7 million per second on a day in 2018 when the company's stock price plummeted, and he's no worse for the wear. In some jobs, employees are "nickel and dimed," to use Barbara Ehrenreich's famous book title. Only people who wait tables, clean homes and hotel rooms, or do construction as day laborers can fathom what life on such a pittance means. People in those kinds of jobs could work three lifetimes and not make as much as Zuckerberg lost in one second. Some jobs open up clear career ladders and pathways to advancement; others trap employees in dead ends. Some jobs are secure, with long-term contracts, health insurance, and pension benefits; others are temporary with no benefits except for employer contributions to Social Security. Independent contractors such as Uber drivers don't even get Social Security contributions. Some jobs provide scope for creativity and autonomy; others treat workers as robots, telling them exactly what to do, how fast to do it, and when they can go to the toilet.

A job is not a job is not a job. Like Dr. Seuss with his infinitely varied fish, the jobs number lumps together all these different ways of earning money and counts them as the same. The monthly jobs number is one big metaphor. When the Bureau of Labor Statistics says 201,000 new jobs were created last month, we have no idea how many were stifling jobs that grind people down and how many were interesting jobs that lift people up. To people who have a job or are looking for a new one, the *kind of* job matters almost as much as whether they have one at all.

For someone who's desperate, though, some job is better than none. Maybe we should try to count *unemployment*. Not having a job is pretty cut-and-dried, right? You either have a job or you don't. Not so fast. Here's how the U.S. government counts unemployment. At the behest of the Bureau of Labor Statistics, the Census Bureau surveys 60,000 households every month and asks the head of the household whether each adult member of the household is currently working and, if not, why not. Using the answers to a series of questions, the surveyors divide people into three groups: those who are currently working for pay (employed), those who "participate in the labor force" but are not working (unemployed), and those who are not even in the labor force (chopped liver). Only the middle category counts as unemployed.

If you're getting confused, you're not alone. You've mastered the art of counting triangles, but now you need to learn some rules about counting unemployment.

Lesson 1: If people tell you they're not working, ask them if they have looked for work in the last month. Even if they tell you they've been desperately looking, don't count them as unemployed unless they utter some magic words about *how* they searched. If they tell you they answered a job ad or knocked on an employer's door, count them as "in the labor force and unemployed." If they tell you they study the help-wanted ads every day or spend hours networking to get job leads, place them outside the labor force and don't count them as unemployed. (Don't ask me why. Go ask your mother.)

Lesson 2: If people tell you they're not working but they're looking really hard, ask if they could start a new job right away if they got an offer. If they say "no," no matter what the reason—injury, illness, family responsibilities—count them as "not in the labor force" because they're not "available for work." Don't count them as unemployed.

Lesson 3: People who have been laid off, aren't currently working, and have given up job hunting should not be counted as "in the labor force." They're not even trying to work. Do not count them as unemployed. You may tell them, by way of consolation, that they will be labeled and tallied as "discouraged workers," but they probably won't make the nightly news when the jobs report comes out.

"I've stopped looking for work, which, I believe, helps the economic numbers."

Are you catching on? The "labor force" as created by the Bureau of Labor Statistics is a land riddled with rabbit holes. Lots of people disappear down them and won't be counted as either working or unemployed.

It's not so simple to count employment and unemployment after all. Many strands of homespun philosophy weave through the official rules, the main one being "Money makes the world go 'round." The government decided a long time ago to count only work for pay, with one notable exception: unpaid work in a profitable family business. So, if you volunteer your time on a profitable family farm—raising pigs, for example—you'll avoid the rabbit hole and be counted as employed, even though you're unpaid. But if you volunteer your time in the family home raising kids who might someday turn out to be a Mark Zuckerberg or a Mother Teresa, down the rabbit hole you go, out of the labor force, neither employed nor unemployed.

Excluding almost all unpaid work from the unemployment measure is fair enough. The measure was always intended to capture whether people are able to earn a living. As a gauge of contributions to the economy, though, the unemployment measure leaves some important things out. The human species would wither if no one cared for children and the sick. Churches, schools, and sports clubs would collapse without volunteers. Elections run mainly on volunteer work. Wikipedia wouldn't exist without it. Can work be productive even if it's unpaid? That's a big question that measurers must decide. So far, they say no.

Another philosophical strand runs through the unemployment measure—the notion of the "deserving poor." In 1878, Carroll Wright commissioned one of the first U.S.

unemployment surveys, which he billed as the first official attempt "to ascertain the facts." The country was still in a depression following a financial panic in 1873 and the press was reporting some scary unemployment numbers. As head of the Massachusetts Bureau of Labor Statistics, Wright wanted to refute the high numbers in order to dampen political unrest. He asked town assessors and police officers to count the out-of-work population in their areas and gave them specific instructions about who to count: not people under 18, not women, and only those able-bodied men "who really wanted to work." How the surveyors were supposed to determine "really want" is anybody's guess, but we can imagine that some juicy gossip and rough-and-ready stereotypes went into the mix. Child labor was common at the time and children and teenagers were major contributors to family income, as were women. Wright's counting rules disappeared a lot of workers. He got the "gratifying" numbers he wanted.

Wright would soon go on to become head of the national Bureau of Labor Statistics. His mode of counting unemployment by inquiring into each person's willingness to work became the essential test for unemployment statistics unto today. Two future presidents confronted with the Great Depression, Herbert Hoover and Franklin Roosevelt, would borrow Wright's strategy: first, talk about how unscientific the unemployment numbers are, then insist on counting only deserving people. As the president of the American Statistical Association warned, statistics are "always accessory to some purpose." If you want to make sense of a statistic, track down the purpose it aided and abetted.

At a rally during the time between the 2016 election and Donald Trump's inauguration, Trump called the government's unemployment number "totally fiction." He elaborated on his claim with a story much like one you just learned in your job training at the Bureau of Labor Statistics: "If you look for a job for six months, and then you give up, they consider you give up. You just give up. You go home. You say, 'Darling, I can't get a job.' They consider you statistically employed. It's not the way. But don't worry about it because it's going to take care of itself pretty quickly."

Critics accuse Trump of disregarding facts and making up his own truths. In this case, though, Trump had a certain kind of truth on his side. The unemployment number *is* a fiction in two important senses. To begin with, its authors have selected only a few kinds of unemployment to include in their count. Counting unemployment is just a more complex version of how we teach children to count cookies: look for one or two features they all share and ignore the differences. The unemployment number is a giant metaphor.

The unemployment number is fiction in another sense, too. It doesn't measure what people actually do. It measures what people *say* they do when they answer survey questions. Some people are ashamed they don't have a job and won't admit it to an interviewer. Some people work off the books and won't admit that either. The unemployment number captures people's stories, not their actions. Even worse for accuracy, people don't get to tell their stories in their own words. Census Bureau employees shape the stories by eliciting only the bits of information that interest their bosses. The unemployment number is a story told by a government with a point of view.

That said, however, the unemployment number isn't "totally" fiction. It's anchored in real situations. It comes from an earnest attempt to measure a social problem scientifically, even if it incorporates some particular views about what should and shouldn't count as unemployment. Once the statisticians have set forth their definitions of "unemployed," they stay true to the definitions while they count people. They don't randomly stick people into categories. Before they do their surveys, they test the wording of their questions on focus groups to ensure that, as much as possible, respondents will interpret the questions in the same way. Moreover, the government does tabulate the number of discouraged workers, the people who Trump says have given up. That unemployment rate goes by the geeky name U-7 and is published every month, too. But geeky U-7 is no crowd-pleaser and no match for Trump's evocative story about a man who slumps home to tell his honey that he hit a brick wall.

Trump knows that statistics carry stories about people's lives and he knows how to compose fiction to make statistics come alive. He challenged the validity of the unemployment numbers by speaking to an emotional truth. Many people in Trump's audience were no doubt discouraged workers who felt threatened by poverty and loss of purpose. Trump tapped into their fear and told them that the government doesn't care about them. That officials dismiss their suffering by calling them "statistically employed." That people like them don't count.

This book explores how counting works in the social world and why numbers can't do everything we wish they could. Numbers have come to serve as reality tests ("If you can't measure it, it doesn't exist") and truth detectors ("Show me the numbers"). Too often people rely on numbers to make and justify their decisions, instead of doing the hard work of thinking, questioning, and discussing. If I devote more space to unpacking the limits of numbers than to praising their virtues, that's because I hope to make them serve us better.

Done right, counting can help us think more clearly about what we value and what we mean by our big words. Words such as "equality," "democracy," "poverty," "danger," "peace," or "productivity"—those are powerful ideas but feeble instruments. Their very fuzziness inspires many a book, film, and seminar. The exercise of trying to measure vague concepts like these forces us to ask ourselves what they mean, or rather, what *we* mean by them. Counting sharpens our minds if we count mindfully.

As a university professor, I feel my mind getting sharpened every time I grade a student paper. I'm forced to measure fuzzy ideals. What makes a good paper? What is good thinking? All teachers know that our grading is partly subjective, even as we aspire to be scrupulously objective. Part of my evaluation hinges on whether the student got the facts right. Does the student understand the lectures and the reading material and convey their content accurately? Part of my evaluation hinges on more subjective factors: Can the student grasp big ideas and theories beyond small facts? Apply big ideas to new topics? Does the student ask original questions or put forth a novel way of seeing the topic? Once I've answered these questions, I have to ask myself how I weigh the different factors. What if the student makes an elegant argument but the work contains a whopping error?

Whether I'm putting a letter grade or a number grade on a student's paper, having to categorize the work forces me to question myself and articulate my reasons for deciding as I do. I'm always aware that my students' hopes and self-image are riding on my decisions, and that each student deserves to question me about why I gave the grade I did. As I'm thinking about a grade, I imagine how I'll justify it to a disappointed student. If I think students deserve a high grade, I justify that, too, by telling them what they did well.

To nail the point, I'm not proposing to do away with counting (as if we could). I'm asking that while we count, we think about the good or the damage our numbers could do. Our numbers will serve us better if we reflect on how we arrive at them as carefully as we hope others do when they make judgments about us.

2

How a Number Comes to Be

The great child-development pioneer Jean Piaget devised an experiment to find out whether young children can count. He lined up two rows of 6 clay pellets and asked three- and four-year-olds whether the rows were the same or one row had more. Most children thought the rows contained the same number of pellets. While the children watched, Piaget removed 2 pellets from one of the rows and stretched it out so it was longer than the row with 6 pellets. He asked again if the rows were the same or one had more. Now most of the children thought the longer row had more pellets. Piaget concluded that toddlers don't understand quantity.

Several years later, two MIT psychologists repeated Piaget's experiments, only this time they used M&M's instead of clay pellets. They didn't ask the children which row had more. They offered, "Take the row you want to eat and eat all the M&M's in that row." Most children reached for the row with more M&M's. "Given sufficient motivation," the psychologists concluded, children can overcome their counting difficulties. Indeed, motivation explains a lot about counting. The correct answer to the question "Which group has more?" is "It depends on whether I want to eat it."

Humans aren't the only creatures who count. Birds do it. Bees do it. Rats do it. Even fish do it. Guppies given a choice between joining a small school of fish and a larger one choose the larger one. They have enough number sense to know where safety lies. Most experiments with animals use food as a reward for getting the right answer. Pigeons and rats learn to count in order to eat. Remember the statistician who advised, "Numbers are always accessory to a purpose." Even animals count for a purpose.

Numbers are the answers people get when they want to eat what they count. Or, when they don't want to be eaten by those who count *them*. If we want to know how numbers materialize, we need to know more than what rules people use to classify things they count. We need to know what motivates them to count in the first place. What do they notice about their world that they think is worth counting? Why do they care about the numbers they'll get when they're done counting?

A famous experiment you can watch on YouTube shows a video of people playing basketball and asks viewers to count how many times the ball is passed. During the video, a person dressed in a gorilla costume strolls through the fast-moving tangle of players. When the video ends, the researchers ask viewers if they noticed anything unusual. Very few people noticed the gorilla because they kept their eyes on the ball, exactly as they were told. The experiment demonstrates what the researchers call "inattentional blindness": we don't see what we aren't paying attention to. And, I'll add, if we don't see it, we won't count it.

Health researchers have measured how breastfeeding and formula feeding affect babies' and mothers' health, but in all the scientific hoopla, no one paid much attention to Dad. I wouldn't have thought of taking a father's point of view, either, until I read an essay by Nathaniel Popper about his family's experience with the breast-versus-bottle debate. Popper and his wife had read all the studies suggesting that breastfed babies are less likely to develop asthma, diabetes, or earaches, and that nursing moms have a lower risk of breast cancer. His wife went to great lengths to breastfeed their first son but it wasn't happening, so she reluctantly gave in to formula.

The switch to formula, Popper writes, "gave my relationship to my son a depth that I, as a father, would have otherwise missed out on, and that has continued long after he stopped drinking from a bottle." When their son cried at night or in public, "I instinctively started toward him." Before formula, his wife had been the "first responder" because they both assumed crying meant the baby was hungry. "Now, I was just as capable of feeding him as she was. This meant that I not only fed him but learned about all the times when he wasn't actually hungry but needed a burp or a clean diaper, or something else we couldn't figure out but was part of the essential mystery of parenting." When their second son came along, Popper secretly hoped that breastfeeding wouldn't work so he wouldn't miss out on "those endless hours of providing my baby with exactly what he needed."

Fathers are invisible gorillas in the baby-feeding debates. Few studies, if any, try to count the impact of breast- and bottle-feeding on fathers' experience of parenthood, their relationships with their children, the parents' relationship with each other, or the division of work and responsibility. "Inattentional blindness" in this case isn't simply a matter of not looking. It comes from deep cultural norms about who's responsible for babies.

Other kinds of blindness seem to be rooted in how our brains process information. We can't possibly keep track of everything happening around us, so the brain takes mental shortcuts. Shortcuts enable us to make rough judgments, but often they distort our counting. Daniel Kahneman and Amos Tversky launched a new field of behavioral psychology with clever studies of these distortions, which they called "cognitive biases." Kahneman and Tversky were especially interested in why we often misjudge the frequency of events, such as how often we do the dishes. In one study, researchers asked pairs of spouses to say what percentage of various housework tasks each one did and what portion of the couple's social events each one initiated. Miraculously, the spousal contributions added up to more than 100 percent. No doubt each spouse remembered more of his or her own efforts than the spouse's because we pay more attention when we're the one sweating over chores.

People are strongly influenced in their perceptions of how often an event occurs, Kahneman and Tversky found, by how easily they can recall instances of it. This quirk became known as an "availability bias." I'm guessing they chose the word "availability" to suggest that when things are in the forefront of our minds, for whatever reason, they're more available to our thinking brains—the part that does the counting.

So the next question is, What shoves things into the forefront of our brains? Certainly, we all know the power of suggestion. If someone tells you to watch the ball and count the passes, they're suggesting what should occupy the front of your brain. News stories put things in the front of our consciousness, especially when they're repeated over and over during a short time span. News or events we've seen recently move to the front of our awareness and displace more distant events. Our emotions influence what sticks in our brains, too. Frightening or exhilarating stories get our juices going and will hold our attention longer than news about the water department's change in office hours.

Which do you think kills more people—tornadoes or asthma? Asthma kills 20 times more people than tornadoes, but most people think tornadoes are the bigger cause of death. That's because every tornado makes the local news, but few cases of asthma do. Every news story about a tornado includes any deaths it might have caused. Even if there was only one death, the association between tornadoes and death sticks in your mind. Most discussions of asthma in news or personal conversations are more likely to mention inhalers than death. Unless you know someone who died of asthma, it's not a scary disease to you. Even if you've never experienced a tornado, you've seen videos that scare the daylights out of you—or watched *The Wizard of Oz*. All these factors make you store deaths by tornado in the front of your brain, but not deaths by asthma.

Another way the power of suggestion works is by firmly anchoring an idea in our minds so that we keep dwelling on it, just as a boat's anchor keeps it circling around one spot on a lake bottom. Anchoring effects are common when people are buying and selling things and haggling over the price. Here's an example. Real estate agents price a home by weighing its location, size, layout, quality of workmanship, and special features. They identify comparable houses nearby that have recently sold to come up with a fair market value. That, at least, is how the real estate market is supposed to work. One study aimed to find out what factors most strongly influence agents when they estimate a home's value. The researchers invited agents to visit a house and gave them an information packet that included, besides all the relevant features of the home, the asking price. Unbeknownst to the agents, half the booklets had a fake asking price that was substantially higher than the actual asking price, while the other half had a fake asking price that was substantially lower. At the end of their house tour, the agents were asked what they thought would be a reasonable price for the house. They were also asked to say what factors had influenced their thinking. None of them mentioned the asking price, but clearly, those prices influenced their thinking. The group that saw the higher asking price gave higher "reasonable" prices than the group that saw the lower asking price.

"About my allowance, care to throw some numbers around?"

When it comes to how we count, our motivations are even more powerful influences than cultural blinders and perception quirks. Sometimes how many we see depends on what we want to do about our findings. In 1994, genocide in Rwanda pushed thousands of Tutsi refugees to flee to neighboring Zaire. The U.S. didn't want to get involved and certainly didn't want to send troops. Several European countries and the United Nations High Commissioner for Refugees wanted to send troops and tried to pressure the U.S. to join in a rescue. The two sides duked it out by invoking counts of refugees. A large number of refugees would argue for intervention; a small number would prove that sending troops was unnecessary.

Needless to say, the U.S. disputed the UN's high numbers. To break the impasse, humanitarian organizations called for examining satellite images and aerial photos of refugee camps in Zaire as objective evidence. However, a satellite photo can't show individual people; at best it shows clusters of people. When the two sides examined the photos, they interpreted the images differently, in accord with their predispositions. UN officials estimated 750,000 refugees among the clusters. U.S. officials saw almost none and politicians back home congratulated themselves for not wasting American military effort. The two sides' goals and motivations explain why they got different numbers. They saw what they wanted to see.

In many situations, people try to motivate other people to change their behavior by giving them incentives—rewards if you do what I want, penalties if you don't. Parents, teachers, employers, managers, negotiators, and diplomats all use incentives to herd their charges into line and guide them toward the same goal. Rewards and penalties often get results. They motivate people to change their behavior, but they can lead to some seriously distorted counting. If someone hires you to count her wares and offers to keep you supplied with M&M's so long as you come up with numbers she likes, you know what to do. The more you want or need the reward, the more you'll be tempted to bend the numbers. This kind of temptation led to the financial crisis of 2008—and we're no longer dealing with itty-bitty candies.

In the run-up to 2008, banks had provided mortgage loans to people who couldn't possibly have paid them back. Wall Street firms bundled shaky mortgage loans into large bonds, where the riskiness of the original loans was less visible. When you buy a bond, you're loaning money to a company or agency that issues it. The bond is a promise to pay you back the full amount of your loan plus interest. Bond-rating companies, such as Moody's and Standard & Poor's, measure the likelihood that a bond issuer will be able to keep its promises. The ratings are expressed in letters similar to school letter grades—for example, AAA, AA, A, B+. The grades tell investors how safe a bond is, just as academic course grades supposedly tell employers and grad schools how competent and creative a student is.

But here's the hitch: bond-rating companies get paid by the bond issuers to rate their bonds. If a rating agency wants future business from a client, it had better rate the client's bond favorably. So when powerful Wall Street firms paid to have their mortgage bonds rated, the rating agencies happily gave them good grades. In education, such a pay-for-grades scheme is called bribery. (As in: wealthy parents pay SAT test proctors to doctor their children's scores.) In business schools, this kind of coziness might be called good customer service—giving your customers the quality they want. In arithmetic, it's called letting your self-interest influence how you count. In the real world, such schemes can lead to disasters like the financial crisis.

The moral of this story? If we want to know how a number comes to be, we need to know that people count in a different way, and not altogether independently, when their rewards hinge on their numbers.

When rewards and penalties are essential to people's livelihoods, they feel more pressure to make their numbers look good. People can be marvelously inventive without engaging in deliberate lying. Creative counting could be a fitting name for what happened to welfare programs after President Clinton's big welfare reform. Since the reform, states measure success of their welfare programs primarily by how many clients find jobs. Many states contracted out their income-assistance programs to for-profit firms and began paying them for success. Firms that don't meet states' targets face pay cuts or, in the worst cases, losing their contracts. From the program managers down to the caseworkers, everyone focused on "what our numbers need to be." Caseworkers pushed clients into taking any job, no matter how low-paying or stultifying, so they could be counted as employed. Caseworkers increased the number of successes by counting clients who took steps to build their "job readiness" skills even though they didn't find jobs. Merely showing up for appointments with a caseworker and turning in the proper paperwork counted as "successful transition to employment."

Creative counting isn't lying with statistics by making them up. It's deception by finagling with the first step of counting: categorizing what counts as a what.

Sometime in the 2000s, the United Nations General Assembly decided to get serious about gender violence, so it did what all political bodies do when they decide to get serious. It set up expert committees and told them to come up with a solution. In this case, the committees were told to develop indicators to measure levels of violence against women in different countries. Indicators break down a fuzzy concept into small events that people can count. For gender violence, the events would be behaviors that people consider to be physical, sexual, psychological, or economic violence. Eventually, surveyors would go around asking women in each country, "Have you experienced this or that behavior?" Add up the "yesses" and we've got our measure of violence against women.

The general idea seemed workable enough, but decision-by-committee is never simple. There were meetings and more meetings attended by a shifting cast of characters from universities, UN member countries, and agencies concerned with women's issues. Committee members quickly found themselves tossing around hot potatoes. Should female genital mutilation count as a form of violence against women? The first mention of it made the attendees aware that maybe they ought to invite someone from Arab and Muslim countries to be in on this discussion. What about child marriage? Does that count as violence? Uh-oh—countries define the age of adulthood differently.

At a meeting in 2009, six years after the General Assembly's first call for numerical indicators, representatives from Europe, North America, Australia, and New Zealand put forth their ideas about violence, based on gobs of feminist writing and victim surveys in their own countries. Their ideas about physical violence included hitting, kicking, biting, slapping, pushing, grabbing, shoving, beating, and choking. Women from the Global South didn't have as much research to draw upon, but they did have plenty of experience with violence. Some Bangladeshi women proposed adding other acts of physical violence to the survey: "burning," "throwing acid," "dropping from a high place," "smashing the hand," and "needling the finger." They also proposed some forms of psychological violence that hadn't occurred to the Western experts. Here are three of them, with some hints about what they mean:

"Expelling women from the house": In some parts of rural Asia, women aren't allowed inside the house while they're menstruating and for three months around childbirth. During these times, they're forced to sleep outdoors in animal sheds, or on the bare ground if their family has no animal shelters.

"Rebuking women for giving birth to a girl": In many parts of Asia, all status and power belong to men. A woman who doesn't produce a male child might be rejected by her husband and his family, beaten, starved, and driven out of their village. A Pakistani woman I know was married off at a very young age to a man who turned out to be impotent. To cover his humiliation, he beat her. His family took his side and ostracized her. When she eventually fled back to her parents, they discouraged her from seeking divorce because it would bring shame on them, too. By the time I learned her story, she was a PhD student at Brandeis University.

"Taking additional wives": We're not just talking affairs or serial marriages here. An Afghan human rights activist I know recalls what it was like being the daughter of her father's first wife. When an Afghan man takes an additional wife, previous wives and their children lose dignity, affection, and influence. My friend's mother became in essence a servant to the man and his second wife. It was bad enough to be female in this patriarchal, polygamous society. Now she was a *third-class* female.

The United Nations "Guidelines for Producing Statistics on Violence against Women" that finally emerged in 2014 didn't include the behaviors that the Bangladeshi women considered physical and psychological violence. When the surveys are complete and the numbers come in, they'll reveal little about the people whose ideas weren't included in the heap of things to be counted. The next time you hear someone say "stories aren't data," think about these women, because their stories *should be* in the data.

Scientists have a word for a measure that captures reality and measures what they want it to measure: validity. They know that methods of collecting data can produce deceptive numbers and so they strive to make their numbers valid. But there are no clear rules for judging validity, nothing like "If it has 3 points, it's a triangle." Textbooks on research methods warn young scientists to identify "threats to validity" and to eliminate them as much as possible, but the textbooks don't tell them how.

The best way I know to find out whether a measure is valid is to travel back to the moment of creation and nose around. Who was in the room where it happened? Who was asking the questions and what did they think to ask about? Who got to say what "counts as" the thing being counted? With the UN's gender-violence measures, Bangladeshi women were in the room, but their ideas weren't included in the final measure. Being in the room isn't always enough, so here's another way to find out whether a measure accords with reality: ask the people whose lives you're measuring and those who will be affected by the numbers your measure yields. When they tell you something, listen to them, take them seriously, and include their voices in your measure.

One way to get expert consensus on how to count is to hold meetings, as the UN gender-violence project did. Put experts together in a room and let them discuss, debate, negotiate, and reach a consensus. As we all know from experience, though, small groups can easily be dominated by one or two people. With the help of computers, there's a new way to reach expert consensus. Computers can process thousands of judgments made by experts sitting alone at their desks, undisturbed by other experts.

Political scientists have developed a sophisticated expert-consensus technology to measure democracy, in hopes of learning why some democracies are more robust than others. Varieties of Democracy, or V-Dem for short, is a project based at the University of Gothenburg in Sweden. The project relies on 3,000 experts to rate democracy in 200 countries. For each country, five experts answer hundreds of questions in an online survey. For example, here's a question about gender equality:

> **Question:** *Do women enjoy the same level of civil liberties as men?*
> Clarification: Here, civil liberties are understood to include access to justice, private property rights, freedom of movement, and freedom from forced labor.
>
> **Responses:**
>
> 0: Women enjoy much fewer civil liberties than men.
> 1: Women enjoy substantially fewer civil liberties than men.

2. Women enjoy moderately fewer civil liberties than men.
3. Women enjoy slightly fewer civil liberties than men.
4. Women enjoy the same level of civil liberties as men.

On each question, experts rate their country on a scale from 0 to 4. With a lot of mathematical wizardry, V-Dem combines the expert answers into scores on a 1,000-point scale, running from 0 to 0.001 and on up through 0.999 to 1.00. No country gets a perfect 1.00, but Norway leads the pack with 0.867. No one flunks with a 0, but North Korea comes close with 0.010.

Since V-Dem's numbers come from a computer-generated blend of individual expert opinions on hundreds of questions, I wanted to go back to the moment of creation, so to speak, to learn how the individual experts' numbers come to be. I asked a colleague who serves as a country expert for V-Dem to talk with me about her experience. First, I asked her to tell me how the survey process works. She told me the experts get two weeks to answer the survey and she allowed as how she couldn't always give it her undivided attention. "I do it in a couple of nights. But I'm aware that I've got other things to do. You're in a hurry, it's due tomorrow, I've still gotta prepare my class and it's nine at night."

Right, I thought to myself. You can't take the human out of the expert. Worth remembering, even though my colleague is dedicated to high standards of research. For all the imperfections of the measure, she emphasized, she finds the ratings very useful for her research and teaching. "If I want to ask why democracy is faring better in some countries than others, if I don't have some kind of measure of democracy, I can't begin."

I had selected three questions from the survey and asked my colleague to walk me through her thought process as she answered them. I started with the question about gender equality and civil liberties, because when I first read the question, I was totally baffled. The term "civil liberties" is "clarified" by yet more vague concepts: justice, rights, freedom, property, and force. The subtle differences among the terms "much fewer," "substantially fewer," and "moderately fewer" left me scratching my head. I wanted to understand how an expert deals with these ambiguities to come up with a numerical rating.

Me: "When you think about this question, do specific events or concrete examples go through your head?"

She: "Yes." Her reply was immediate and emphatic.

Chalk one up for V-Dem, I thought. She's not answering on the basis of vague impressions. She's bringing detailed knowledge to bear. Then I explored how she deals with the ambiguities in the question.

Me: "Suppose you think women in your country enjoy about the same level of freedom from forced labor as men (a 4) but much less access to justice (a 0). How do you decide which of the numbers to give? Would you choose a 2 because it's in the middle? Or might you choose a 1 or a 3 depending on whether forced labor or access to justice matters more to you?"

She: "My answers are not that precise. They're more impressionistic. You'd have to be doing a PhD dissertation on the status of women in [my country] to rigorously answer that question. I can't think of anyone in the U.S. who would have a better ability to answer that question [than I], but no way I really have the knowledge to answer it."

For the two other questions I had chosen, my colleague had much more specific detail at her fingertips to back up her score. She articulated her thought process so clearly and convincingly that I came away thinking that the V-Dem survey does in fact draw on deep knowledge. For all the human foibles that afflict experts while they answer the questions, their numbers rest on some solid factual detail and on their earnest desires to advance knowledge. Yet even someone as statistics-savvy as my colleague sees the impressionistic nature of the counting enterprise and the false precision of a 1,000-point scale. Each number results from an assemblage of short, sometimes hurried, ruminations about issues on which the experts have varying degrees of knowledge.

Yet, as I thought more about the gender-equality question, I encountered a problem much deeper than ambiguity or false precision. Suppose you think one's body is the most valuable private property a person can own, or that being prevented from ending an unwanted pregnancy and required to raise a child is a kind of forced labor. The question doesn't allow you to express your ideas about what the terms "private property" and "forced labor" mean. If your country or the state where you live outlaws abortion or, like China until recently, forces pregnant women to have abortions after they've already had one child, how do you answer this question?

If you downgrade your country because of its laws on reproduction, you have no way to tell the V-Dem scientists why you downgraded. Worse, if you weigh reproductive freedom more heavily in your rating than your fellow experts do, your answer will be discounted. It may seem odd to discount experts whose judgments aren't in the mainstream. This happens because V-Dem works hard to make sure its numbers are *reliable*. For scientists, reliability means that when different people count the same thing, they get the same result. V-Dem's scientists know that experts have different thresholds for classifying answers on the 0-to-4 scale. To correct for this variety, V-Dem applies some fancy math to the experts' answers to determine "the degree to which they agree with other experts." And then the kicker: "Experts with higher reliability have a greater influence" on the final scores.

You read that right. The more you go along with the crowd, the more influence you have on the final score. It's a paradoxical power, though, for if you think the same way as the crowd, the crowd probably has had more influence on you than you will have on its conclusions. Because scientists define reliability as counting the same way other trained scientists count, conformity becomes the standard of correctness. To invoke one of my mother's mantras, "Just because everybody does it, doesn't make it right." The method produces a kind of groupthink, albeit with experts instead of average Joes.

Some numbers come to be through a kind of bucket brigade. In the days before fire engines and tanker trucks, people formed a human chain to pass buckets of water from a lake to the fire. This is how foreign-aid agencies get information about economic progress in developing countries. Local agencies gather economic data, then pass the figures up to state officials, who pass them up to regional agencies and on up the data go to national ministries and statistical agencies. The national agencies combine the numbers and pass on the totals to international-development agencies, especially the World Bank, which bills itself as a "knowledge bank."

The numbers at the top of the bureaucratic data chain, the ones that make it into the knowledge vault, are only as accurate as what comes up from the bottom. Local statistical agencies are notoriously underfunded, understaffed, and underequipped. How much livestock do farmers have? The answer depends on how much money *you* have to count it. Livestock numbers in Uganda went up 60 percent after its agricultural census received extra funding. (Not to single out Uganda here: The U.S., the Census Bureau has been the victim of partisan budget politics since the nineteenth century. Political parties and interest groups who want more accurate and inclusive population counts feed the bureau money. Those who would rather not see much change from previous censuses starve it.)

Agencies all along the data chain would like to have numbers that are often hard to get or nonexistent. So what do they do about missing data? Sometimes they make assumptions and estimate. Sometimes they just plain make up the numbers. And sometimes it's hard to draw a line between estimating and "making it up."

In developing countries, subsistence farming is a big part of agriculture. People eat what they produce and barter for what they can't grow themselves. Little or none of their produce enters the market, where its value could be tracked by firms and administrative agencies. A lot of other economic activity takes place informally, off the radar of the tax man. Many African countries estimate the value of subsistence farming by making a happy assumption. They assume annual food production grows at the same rate as rural population. That assumption builds agricultural progress right into the measure. Why bother counting if you already know the answer?

At the low end of the bureaucratic data chain, employees in local agencies sometimes make up the data. They're required to submit annual reports to national agencies but not to document where their numbers come from. One civil servant in an African agency allowed as how he routinely reports crop production as 10 percent higher than he reported the previous year.

Fudging happens the world over in small organizations and large, as you have undoubtedly observed and may have practiced yourself. People fudge for many reasons, not least because paperwork is a bother. Civil servants, doctors, teachers, and business owners feel that paperwork isn't their primary task. They didn't spend all that time in school to learn how to fill out forms. Enumerating to please superiors or outside agencies is something to be dispatched quickly so they can get on with the real work. One of my colleagues wanted to determine the most common causes of industrial fires. She thought she could get good numbers from the incident reports that firefighters file after a fire. Her hopes were dashed when a fire official told her, "We fill it out on the way [to the fire] because it wastes time if we do it when we get there, so we just kinda put something. Sometimes we just make it up because we have to submit the form to our chief."

Climbing back up to the top of the development data chain, the World Bank fills in missing data with its own "method of gap filling." Say the bank wants to know how much maize production increased in Country X from 2000 to 2010 but there are no data after 2003. With the gap-filling method, the bank assumes that maize production in Country X during the missing years increased as much as it did in nearby countries at similar levels of development. In case you didn't follow that, the World Bank makes up missing data for Country X by lending it other countries' results.

Estimate is a fancy word for guess. If we have no idea how many there are, we guess by taking a shot in the dark. If we have some understanding of a situation, we guess by making assumptions. Assumptions are a form of fiction. "Let's assume . . ." is the logician's way of saying, "Let's pretend . . ." without sounding like he's telling a fairy tale. Any number that's based on assumptions is part fiction. When using estimated numbers, we shouldn't forget where they come from or let ourselves be taken in by fantasy.

That's not to say we shouldn't ever try to estimate when we don't have complete data, but some techniques for estimating are more tightly tethered to reality than others. Let's take a course in Bird Counting 101 from the Cornell Lab of Ornithology. Environmental scientists use species counts to track a population's health and changes in its breeding and migration patterns, perhaps in response to changes in its habitats. Cornell's eBird website gathers bird counts from amateur birders all over the country. It's easy enough to count 8 or 10

Goldfinches perched on trees in your yard, even if they're flitting around. Counting Crested Auklets in a large, swiftly moving flock over the ocean presents a challenge.

The Cornell ornithologists suggest how to break down a large counting problem into small increments. In the flock of Crested Auklets, you count the first 10 birds in the flock (as shown inside the white circle), make a rough guess at how much space they take up within the flock, then eyeball how many circles that size you can "see" in the whole flock. Take a minute to try this method yourself.

When I tried this exercise, I got 100 birds. The experts say they got 140, and the actual count is 118 birds. Not bad. Rough estimates like yours and mine, multiplied by many bird-watchers, are sufficient to help experts interpret whether the species population is small, common, abundant, holding steady from year to year, or in decline.

The difference between 100 and 118 isn't biologically significant. But there's a huge difference between the ornithologists' estimation and the World Bank's "gap-filling method." The World Bank estimates Country X's maize production by starting from the last year for which it has data and then adding other countries' growth rates to it. Beyond the last year with data, the estimate isn't pegged to maize actually grown in Country X. The ornithologists have us start with a quick count of an actual observation, then continue using our own eyes to mentally overlay our circle on the rest of the actual flock. This is what I mean when I say some estimates are more tightly tethered to reality than others.

To return to the language of science, measures are valid if they actually measure what they purport to measure. Estimates based on actual observations of whatever we're counting have greater validity than ones based on assumptions pulled out of the air —if those can be said to be valid at all. An estimate of agricultural crops that is so removed from reality doesn't serve the World Bank's purpose because it can't reveal whether agriculture policy initiatives succeeded or failed.

Any discussion of how numbers come to be has to consider how computers count. Computers *don't* count. People count and enlist computers to help them. Computers help by collecting and storing unimaginably large quantities of data. They register every time someone clicks anywhere on the web, keeping track of humanity through its digital footprints. Computer software can link different data sets together, enabling humans to find out more about any topic or person by pulling together information from different sources. The term "Big Data" is shorthand for all these ways computers help us count when we run out of fingers and toes and our brains don't have enough space or horsepower.

By counting digital footprints, computers can reveal how many people clicked on a website or an ad, how long they looked at it, and where else they went on the web after leaving the site. They can tell you how many Friends your teenage daughter has,

who they are, what they've done recently, and who else they hang out with. They can tell a business how many people Like its products or a Twitter user how many people follow him.

At first glance, it might seem that counting hits, Friends, and Likes yields accurate numbers untainted by incentives and social pressures. People either clicked or they didn't. What's to dispute about that? A moment's reflection suggests what. Businesses offer bonuses to Like their products and to broadcast your Likes to your friends. Think how you search for music or videos on YouTube and decide which ones to try. Your searches turn up a list of titles along with the number of views each one has had. I don't know about you, but try as I might to not be influenced by popularity, I can't stop myself from thinking that a music video with over 100,000 views must have more going for it than one with only 972 views. Sometimes I click on the one with the least views just to be contrarian, but most times I click on the one with the biggest number. I go with the crowd.

Social network sites such as Facebook stir up the same kinds of hormones as any high school cafeteria. People Friend or join networks partly to advertise who they admire and want to be associated with. Accepting people who ask to be your Friend may be a sign of trust, admiration, or affection, but more likely you accept requests because it's socially awkward not to. Site members develop strong rules of etiquette, such as "It's rude not to answer someone's request to be your Friend right away." Your daughter's social network site doesn't tell you anything about why she chose each Friend or whether she actually spends time with them outside the cybersphere. The numbers of hits, Friends, and Likes don't count anything real other than clicks and keystrokes. They're the products of human needs to be accepted and feel connected.

Computers use algorithms to sort through large amounts of data and make them useful. An algorithm is like a recipe, a set of rules and procedures for getting something done. Programmers bundle rules together into algorithms designed to answer human questions, such as "Which ad will net me the most sales?" or "Which job candidate will perform best for my company?" People sometimes use the terms "Big Data" and "algorithms" interchangeably, but to be more precise, Big Data refers to the massive quantity of information floating around in computer databases and on the web. Algorithms are the software programs people use to generate predictions and advice from data.

Algorithms do more than simply record what we do on the web. They influence how we behave on it. Facebook's, Google's, and YouTube's business models are driven by advertising revenue, so these companies direct our attention to sites where (they hope) we'll spend more time watching, reading, and exposing our brains to their advertisers. Facebook feeds us only the news it thinks will interest us and keep us glued. YouTube queues up videos in a sidebar titled "Up Next for You," sparing us the trouble of searching for what interests us. All these friendly suggestions are served to us by "recommender algorithms."

Recommender algorithms have figured out that lots of people are drawn to sensationalism, conspiracy theories, and extremist political views. According to a former Google employee, YouTube's engineers didn't care about the truthfulness of videos on the site. "Watch time was the priority. Everything else was considered a distraction." What does "watch time" actually measure? If you believe a YouTube spokesperson, watch time provides an accurate "reflection of viewer interest". If you believe the Invisible Gorilla principle, watch time measures what catches people's attention when they're fed suggestions by people with a monetary interest in their behavior. So much for objective counting.

Recommender algorithms give us unsolicited advice. Predictive algorithms give us advice when we ask for it. A bank wants to know how it can weed out loan applicants who are likely to default. A firm wants to select the job applicants who will perform best in its company. A city housing department wants to identify substandard buildings at high risk of fires. Those are the kinds of questions predictive algorithms are designed to answer.

Police departments use algorithms to predict where crimes are most likely to occur so they can deploy more police to those areas, who in turn will be able to detect crime before it happens. One of the most widely used—and criticized—algorithms is PredPol (short for Predictive Policing). Like all predictive algorithms, this one uses data about the past to predict the future. The algorithm counts two kinds of data to predict crime: 911 calls to the police, and criminal activity that police on the beat observe or suspect.

As the algorithm counts, a not-so-funny thing happens. Police chiefs send more police to neighborhoods that initially had a lot of 911 calls and arrests. The police arrive in those neighborhoods on heightened alert, having been warned by the algorithm that the crime rate is high. They will inevitably notice more suspicious activity without waiting for 911 calls, in part because the measure told them to expect a lot of crime, and in part because what counts as "suspicious behavior" is highly subjective. Police will err on the side of caution, arresting people who seem the slightest bit suspicious, because their chief sent them on a mission to deter crime.

Data on these new stops and arrests are fed back into the algorithm. In the next round, the algorithm will detect more crime in the "hot spot" neighborhoods where it sent more police to find crime, and fewer crimes in neighborhoods with no extra police. And on it goes, round and around, with the original hot spot neighborhoods producing ever-higher numbers. The predictive algorithm has become a "runaway feedback loop," a vicious circle that can't stop itself.

Look carefully at how this predictive algorithm works. It doesn't predict actual crimes or where crimes will occur. It predicts police behavior. It predicts whether police will arrest more people in a neighborhood where the algorithm told them to go and look for crime. The algorithm influences police to look harder for crime in those places, so indeed, police find more crime in those places, so chiefs send more police to those places. In each round, the algorithm "sees" data that it helped generate. *The algorithm causes the numbers it counts to come into being.* The so-called crime rate isn't a fixed feature of a neighborhood, like its roads and buildings. The crime rate is made by police interacting with an algorithm.

Let's run the same kind of thought experiment on an algorithm designed to help businesses find the best employees. Using the algorithm, an employment agency sends a diverse pool of candidates to an employer. The employer selects more whites, more men, and more people under 50. No one entered race, gender, or age into the algorithm, but over time, the algorithm "sees" that whites, men, and younger people have been more successful for the employer. The agency wants to please its client and the algorithm has been designed to do just that, so it tells the agency to send mostly whites, men, and younger people.

A predictive algorithm is a Hall of Mirrors. A mirror reflects what's in front of it. If the people in front of it behave in biased ways, the mirror will "behave" in the same biased ways. The mirror doesn't know that it's arresting more blacks than whites under the same circumstances or hiring more men than women. It knows only what the police do or what employers do. It assumes *they* know what makes them successful, so it tells them to keep on doing more of the same.

Numbers enjoy an aura of objectivity and precision unwarranted by their origins. They are always products of human judgment, even the numbers that seem to spring from computers untouched by humans.

When we count, we're holding an imaginary clicker in our hand. Lots of things can make us decide to click or not to click. Quirks of perception go into clicking decisions. Motives go into clicking decisions. Culture, tradition, and history go into clicking decisions. Assumptions go into clicking decisions. Once we understand the factors that influence counting, we can start to see how and why all numbers are cooked—not in the sense of faked, but in the sense that every number might be different if these factors weren't in play. And without knowing how a number came to be, we can't know what it means.

3

How We Know What a Number Means

Hans Rosling could eke out half a novel from a 2-digit number. In one of his public health classes, the Swedish professor distributed child mortality tables from a UNICEF report and asked students to read aloud the numbers as he named a country. Saudi Arabia? "Thirty-five" came the answer. "Correct," Rosling said. "This means that 35 children die before their fifth birthday out of every thousand live births." After eliciting numbers for a few countries, he stopped to explain "why I'm obsessed with the numbers for the child mortality rates." Because, Rosling went on, "this measure takes the temperature of a whole society. Because children are very fragile. There are so many things that can kill them. When only 14 children die out of 1,000 in Malaysia, this means the other 986 survive. Their parents and their society managed to protect them from all the dangers that could have killed them: germs, starvation, violence, and so on. The number 14 tells us that most families in Malaysia have enough food, their sewage systems don't leak into their drinking water, they have good access to primary health care, and mothers can read and write."

I'm in awe of someone who can paint such a vivid picture of life in Malaysia from the number 14. I'm also baffled by the disconnect between Rosling's broad-brush paint strokes (enough food, good health care, clean drinking water, literacy) and the pinpoint precision of 14. There's something magical going on. A 2-digit number can't possibly be large enough to contain everything in his painting. Everything in the painting can't possibly be shrunk into the number 14 without losing its vitality.

As a teen, Rosling wanted to learn how to swallow a sword. Not having a sword handy, he practiced with a fishing pole, but he couldn't make it work. A few years later, he was a doctor in training. A professional sword swallower landed in his ward as a patient. Rosling grilled him for the secret. "Young doctor," the man said, "don't you know the throat is flat? You can only slide flat things down there. That is why we use a sword." Sword swallowing is easy if you know some anatomy.

Rosling gives meaning to child mortality numbers with social anatomy. He puts children back into the real-life situations from which the numbers come. He asks himself, as any parent might, What does it take to keep a child alive? Instead of pinning all the responsibility on parents, though, Rosling asks, what does *society* have to do to make it easier for parents to keep their children alive?

Child mortality records aren't pinpoint accurate and their quality varies among countries, but those problems don't detract one iota from our appreciation of Rosling's picture. His verbal sketches are enough to suggest situations anyone can imagine. The most important question to ask about what a number means is the same one Rosling asks: How would people have to live or what would people have to do to produce that number? If you want to decipher accurate meanings of numbers, channel your inner sociologist. Mentally travel among people and find out how they live. A number is only the period at the end of a life story. We don't really know what a number means until someone brings it to life.

Ready for a quiz? Don't be fooled. It looks like arithmetic, but it's really about social anatomy.

Does 3 × 20 equal 2 × 30?

Rebecca Blouwolff, a 20-year veteran French teacher, thinks not. She came to this conclusion when her town school system announced that K-2 students would no longer have their world-language classes 3 times a week for 20 minutes. Instead, they'd get lessons 2 times a week for 30 minutes. She asked herself, What does a language teacher have to do to ensure that pupils learn something? Blouwolff knows the research on second-language learning. "Frequency is essential to language study and outweighs duration." Language learning, she says, "is a case where 3 × 20 does not equal 2 × 30."

Blouwolff breathes life into the school system's arithmetic by putting administrators in her shoes. "As a teacher, I know how often various school events interrupt 'specials' like world language class." If the classes are scheduled only twice a week, "there will be several weeks when one class is missed and students will have language only once a week." She might have written her classroom wisdom as a formula: 2 × 30 minutes per week = 24 minutes per week on average. She might also have suggested another way that 3 × 20 does not equal 2 × 30. It costs more to pay teachers for 3 shorter sessions than 2 longer ones. Sixty minutes aren't always 60 minutes if you understand how people use time.

In Lewis Carroll's *Through the Looking Glass*, young Alice stepped through a mirror and found herself in a baffling, crazy world. There she met a talking egg named Humpty Dumpty, who spoke in nonsensical riddles.

> *"I don't know what you mean by 'glory,'" Alice said.*
>
> *Humpty smiled contemptuously. "Of course you don't till I tell you. I meant 'there's a nice knock-down argument for you!'"*
>
> *"But 'glory' doesn't mean 'a nice knock-down argument,'" Alice objected.*
>
> *"When I use a word," Humpty Dumpty said, "it means just what I choose it to mean—neither more nor less."*

Humpty Dumpty could just as well have been talking about numbers instead of words. Numbers mean exactly what their authors want them to mean. If you have any questions, just ask them—the authors, that is. Don't bother asking the numbers. Contrary to what you may have been told, numbers don't speak for themselves; but their creators, like Humpty, are eager to tell you what they mean. More often than not, numbers are part of somebody's argument.

Every year the Census Bureau publishes poverty thresholds. These are the numbers that determine how many people the government considers poor and how well its welfare policies are working. The thresholds are supposed to reflect how much money a family or household needs to buy adequate food, shelter, clothing, and other necessities. The bureau calculates different amounts of money for different-sized households. Households with income below the threshold count as poor. For example, in 2018 a family of 2 adults and 2 children was counted as poor if its income was less than $25,465 per year. That translates to $2,100 a month.

According to the Census Bureau, "The poverty thresholds can be thought of as the amount of money which, if spent wisely, will be sufficient to meet the basic needs of a family or single person." *If spent wisely*. That's Humpty Dumpty speaking. You're poor if we say so. Any number above the threshold means you aren't poor, or you wouldn't be if you spent your money as we think you should. If you have $25,465 and you're poor, it's your own fault.

The U.S. method for measuring poverty was the brainchild of a lone woman working in the Social Security Administration named Mollie Orshansky. "I don't need a good imagination when I write about the poor. I have a good memory," she told me in an interview in 1993. Orshansky was one of six daughters

"How much do I have to eat to be eligible for dessert?"

of Russian Jewish immigrants to New York. "I wanted to show what it was like not to have *enough* money," she continued. "It's not just that the poor had less money—they didn't have enough. I knew you couldn't spend for one necessity without taking away from another." She knew what happens when the housewife doesn't have enough money for food, because she watched how her mother coped, or, as the Census Bureau puts it, "spent wisely." Her mother served less-nutritious meals. The family often skipped meals. If Orshansky's father brought home a couple of guests for dinner, her mother took the girls aside and told them, "'When the meat comes around, you take one small piece and pass the dish on. Same with the potatoes.' And when the dish came around to Mother, she took nothing." The strategies economists praise as "economizing," Orshansky knew, usually mean doing without.

Orshansky published her first articles about measuring poverty in 1963 and 1964, just as President Lyndon Johnson's War on Poverty was getting underway. In 1965, the Office of Economic Opportunity seized on her method, and it's essentially the same method used today. Before the ink was dry, the critics started harping. Orshansky was first in line at the complaint desk, because the measure the government adopted was far more stringent and less nuanced than the one she designed.

Orshansky used the cost of a nutritionally adequate diet as the starting point for a cost-of-living estimate. Food isn't the only thing a family needs to live, but she chose it as the basis for her measure because the Department of Agriculture had done studies of healthy diets and their cost. When Orshansky began her work in the early 1960s, households were spending about a third of their budget on food. She multiplied the cost of nutritionally adequate food plans by 3 to get a minimum income necessary to keep people above poverty. However, already by 1966, it was clear that people were spending only a quarter of their budget on food. Orshansky tried to get the multiplier raised to 4, to no avail. Today food makes up less than a fifth of the typical household budget, but the government still uses 3 as the multiplier.

The multiplier wasn't Orshansky's biggest concern. The Department of Agriculture had created food plans for 4 budget levels: "liberal," "moderate," "low cost," and "economy." The euphemistically named economy plan, the department warned, was suitable only in emergencies and for short times. Over Orshansky's strong objections, the government used this dire budget for its poverty thresholds.

Orshansky talked back to Humpty Dumpty. She had calculated budget numbers for 124 different kinds of families, so her articles were filled with numbers, but she tried to make policymakers feel what the numbers meant with words. She sprinkled her texts with down-to-earth images of how people get by. According to the economy plan in 1965, a family of 2 adults and 2 children would be just above the poverty threshold if it had an annual income of $3,165. Orshansky broke that number down into the meager 23 cents per meal per person it allowed. Anyone could relate to shopping for a family dinner with 92 cents in her wallet. Orshansky didn't stop at 3 square meals. The economy-plan budget didn't allow for the extras that were no longer extra but had become ordinary parts of middle-class life—snacks, occasional meals out, inviting guests to share a meal, and "money for the ice-cream vendor and soda pop so often a part of our children's lives."

Orshansky's comment about the ice-cream vendor pointed to what is now the main criticism of U.S. poverty measurement. The government uses a fixed definition of poverty—you're poor if you have less than so many dollars. Orshansky understood that poverty is relative. It means lacking not only enough money for a bare-bones standard of living but also enough to feel included in the mainstream way of life. Growing up in a poor immigrant family, she knew the feeling of not belonging.

Most European countries and Canada measure poverty with a relative standard, one that compares how a family is doing relative to everyone else. Relative measures start with the median income. That's the amount of money smack-dab in the middle of the income distribution. If you put that number on the

50-yard line of a football field and ask everyone whose income is above the number to stand on one side of the line and everyone whose income is below the number to stand on the other side, the population would be split exactly in half. Most countries set their poverty level at half the median income. If people don't have at least that much money, they probably can't afford the normal, ordinary things most of their compatriots enjoy.

Economist Shawn Fremstad offers some thought experiments akin to Orshansky's word paintings to help us think about what poverty numbers mean. "If you were a part of a couple raising two children on $25,000 a year or about $2,100 a month, could you afford the basics without going into debt or being evicted?" He might have added that the dollar figure for income doesn't take into account whether people have enough savings to pay for emergency expenses like car repairs and medical bills. Poverty, this question suggests, means not having enough money to get by and worrying that a sudden big expense is going to send you over the edge.

Next, Fremstad asks: If you were part of a couple raising 2 children, "Do you think other people would view you as no longer poor if your family's income was a bit over $25,000?" Poverty, this question suggests, is about more than how you feel. It's also about how you're seen by other people.

Last, Fremstad asks: How much money do you think people need in order not to be poor? This is a question public opinion surveys have asked from the 1950s on. A curious thing comes out of these surveys. As the country's median income rises, most people's estimates of how much money one needs to stay afloat rise, too. To most people, not being poor means being able to keep up with the Joneses. For Mollie Orshansky, it meant having money to buy an ice-cream cone. Today it means being able to afford a cell phone, internet service, a car, and cable TV. Fremstad thinks poverty measures should be redesigned so that the numbers count all these meanings of being poor—not being able to get by, being seen as poor, and not being able to participate in mainstream cultural and social life.

Not everyone agrees that the U.S. poverty measures are too stringent. Some critics think they're too generous, because the measure of family income doesn't include government benefits. If those benefits were counted as income, more people would have incomes above the poverty thresholds and fewer people would count as poor. Since 2010, the Census Bureau has calculated a separate poverty measure that does include some government benefits, such as Social Security pensions, unemployment compensation, and welfare payments. But as economist Robert Samuelson sees it, the new measure still counts people as poor who aren't poor by his lights. Even though the new measure *adds* government benefits into a family's income, it *subtracts* unavoidable expenses such as taxes, out-of-pocket medical costs, work expenses, and court-ordered child-support payments. The Census Bureau doesn't count these unavoidable expenses in a family's income because they reduce the amount of money it can spend on food, shelter, heat, school supplies, or anything else. Samuelson thinks this way of counting income "justifies a lax definition of poverty."

Poverty is many things, but it isn't a number. It's a lived experience, a never-ending discussion about what it means to live a decent life, and an ongoing debate about who deserves help from the rest of society. A country's poverty numbers express how its government gives official meaning to the experience. The critics' arguments express their views of what poverty means and who deserves help.

During October 2018, the so-called migrant caravan increased from 1,400 to 3,000 to 4,500 to 7,200. Depending on where you got your news that month, the caravan was "approaching," "heading, "walking," "traveling," "marching," "trudging," "pouring," "streaming," "thronging," or "surging" toward the U.S. Its members were "trying to cross the border illegally" or "getting ready to invade the U.S." They were "pushing onwards," "defying borders," and "planning a human stampede," though at times, some "lay exhausted, mothers curled up asleep with their children, resting," while others were "dancing and singing."

It's time for another quiz:

Question 1: *Which number is bigger—1,400 or 7,200?*

Question 2: *Which group would you sooner encounter on a dirt road at dusk: 7,200 exhausted people trudging toward you or 1,400 defiant people about to stampede you?*

Question 3: *Which number in Question 2 is a bigger threat?*

I'm betting that the words, not the numbers, drove your answers to Questions 2 and 3. You couldn't answer Question

3 from numbers alone, because what's important in the news stories is not how many people are on the move, but what kind of people they are and what's motivating them. The numbers 1,400 and 7,200 don't mean anything on their own. The one that's bigger in arithmetic may not be the one that's bigger in flesh and blood. It's time to put math aside and turn to literature, because we absorb the meaning of numbers through the stories their authors tell.

One student handbook advises, "When writing about numbers, help your readers see where those numbers fit into the story you are telling—how they answer some question you have raised. A naked number sitting alone and uninterpreted is unlikely to accomplish its purpose." The handbook gets it right except for one thing: Numbers don't have purposes. People do. Writers know how to make us believe that their numbers mean what they want them to mean. They drape their numbers with spiffy verbal attire. Adjectives provide character references: the monthly new job numbers are "steady," "respectable," or "troubling." Adverbs can make us see the glass as half full or half empty. "*Fully* half of registered voters say they support sending troops to [pick your country]." Or "*Only* half of registered voters support sending troops." Verbs and adverbs put numbers on the move. Rarely do numbers simply get bigger or smaller. They swell, they accelerate, and they soar. They inch downward, drop precipitously, or hit a new low. If you describe an increase by saying, "The growth rate exploded," you've planted a bomb in your audience's mind. You've already got them running for the exits. Verbal costumes are the debating tricks, the spin people use to persuade audiences and win arguments.

When authors use numbers to make a point, they carry readers along with plots and stock characters—villains, victims, and heroes. They don't merely tell us how many people are heading toward the U.S. They cast members of the migrant caravan as law-breaking enemies about to wreak havoc or as terrified, beaten-down victims fleeing gangs and machine guns.

Just as there are some standard plot lines in literature—boy-meets-girl, coming-of-age, hero-triumphs-over-adversity, good-guy-vanquishes-bad-guy—there are only a few basic plot lines in number stories. We'll start with the three simplest: More, Less, and Same.

This is more than that.
This is less than that.
This equals that.

Not terribly exciting plots, these. "This is more than that" is a big snooze until you, the reader, put yourself into the story. You have ideas about whether "this" is good or bad and whether "more of this" is desirable or detestable. If "this" is the yield on my investments, I'll take more. If "this" is the likelihood of getting stopped by the police while I'm driving, give me less. "This is more—or less—than that" easily translates into an argument about equality and justice. One percent of families in the U.S. have more wealth than 90 percent of families. (That's not fair!) Blacks and Hispanics receive less screening and treatment for high blood pressure than whites. (That's discrimination!)

Every good plot involves things getting worse before they get better. We can spin variations on More, Less, and Same by adding a time dimension.

> This is getting bigger (growing, increasing, rising).
> This is getting smaller (shrinking, declining, falling).
> This is increasing or decreasing faster than that.

You bring your values to these stories, too. You don't care about size and rates of change in the abstract. You care about these plots when they're about things that affect you. You want things you consider good to grow and you root for people who promise to grow them. You want things you consider bad to decline and you'll do what you can to make them go away.

Because readers interpret stories through their own ideas and emotions, different ways of presenting numbers can evoke different responses. Let's revisit the Census Bureau's predictions that whites will soon be a minority in the U.S. In a clever experiment, researchers asked white people to read news articles about the racial and ethnic makeup of the U.S. They wanted to see how different plot lines would affect the readers' racial attitudes. One group read a story about the proportions of racial and ethnic groups in 2010. The headline was about as unexciting as a headline can be: "U.S. Census Bureau Releases New Estimates of the U.S. Population by Ethnicity." The article told a story of More and Less with percentages and a pie chart showing portions of whites, blacks, Hispanics, Asians, and "other." The second group read a story reporting the Census Bureau's projections of racial and ethnic group proportions in the year 2042. The headline ran, "In a Generation, Ethnic Minorities May Be the U.S. Majority." The article told a story of Some Getting Bigger and Some Getting Smaller. After reading the stories, the people in the experiment rated their comfort level working with people of a different ethnic origin or having their children marry someone of a different ethnic background. Whites who read the changing-proportions story expressed more discomfort than those who read the static description of current proportions.

The two plot lines have different emotional resonance. In democracies, "majority" and "minority" are loaded terms. From preschool on, teachers drill kids on deciding by majority vote. I remember sitting cross-legged on the floor in kindergarten, all of us raising our hands to vote on whether we wanted to hear a story or finger-paint. People who've grown up in a democratic civic culture know in their guts what it means to be in the minority. It means you don't get your way. For whites in the U.S., the racial-shift story says, "Watch out! You won't be Top Dog for long."

The word "minority" has taken on a double meaning that may contribute to the discomfort whites feel with the racial-shift story. Minority in the math sense means fewer than half. In the last few decades, though, minority has become a synonym for nonwhite. To put it bluntly, when referring to race, minority means anyone whites have classified as not white. The word is commonly used to describe *individuals* belonging to racial and ethnic groups that have historically been numerical minorities. Thus, we hear sentences such as "Minority students have a hard time" or "Minorities bring different perspectives to the workplace." Labeling black and brown people as minorities—as in "We want to hire more minorities"—suggests ever so subtly that they are and should remain numerical minorities. It's fine to hire a few of them as long as they don't take over. The language solidifies the political status quo, just as the joists and beams of a building permanently set its shape. With this dual meaning of minority ringing in the ears, the headline "Ethnic Minorities May Soon Be the U.S. Majority" sounds tumultuous and threatening because it's mathematically wrong and democratically wrong.

In the U.S., one more factor heightens the emotional resonance of the racial-shift story for whites. Whites have always been in a position of privilege. At some level, they know their privilege rests on superior power. American-history books teach the archetypal story of the plantation slave revolt, a story of how a politically weak majority (slaves) might one day overpower a politically strong minority (slave owners). It's no wonder that among whites, the story of change evokes fear and defensiveness more than the story of the status quo.

Plots about changing size often come in the form of graphs with lines going up or down. Graphs depict the stories behind numbers with a powerful visual image.

Hans Rosling, the sword swallower who could turn numbers into novels, was on a mission to prove that the world is getting better. Not a small task, but he was undaunted. He filled his lectures and publications with graphs. There are upward-rising lines of good things, such as new movies, women's right to vote, access to electricity and water, literacy, protected land, and scientific publications. There are downward-falling lines of

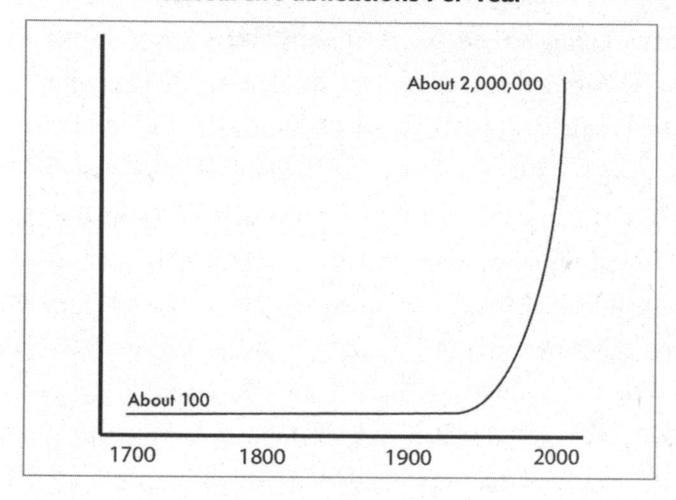

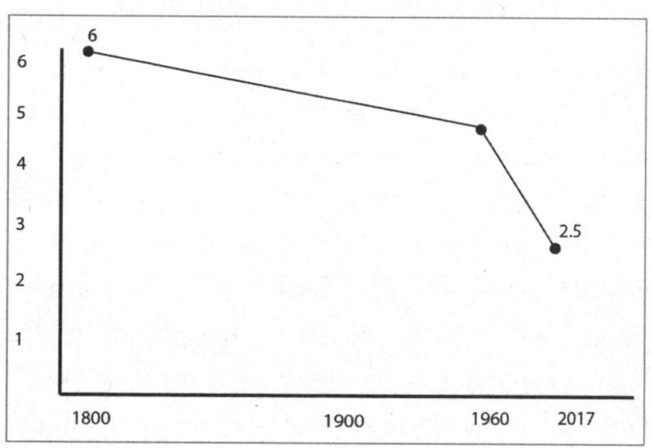

bad things, such as legalized slavery, oil spills, the price of solar panels, child deaths before age five, plane-crash deaths, hunger, and—hold your horses—babies per woman.

At first glance, these graphs are dramatic and convincing. They're convincing because Rosling has told us in words that everything he pictures as rising is good and everything he pictures as falling is bad. Look closely, though, and some of the stories behind the graphs are ambiguous. Not all published research articles improve our lives. According to Marcia Angell, an expert on the pharmaceutical industry, research on new drugs is riddled with conflicts of interest. Most of it is sponsored by drug companies and carried out by researchers who receive financial benefits from drug company sponsors. Many of the published articles about new drugs distort their safety and efficacy, leading the FDA to approve harmful drugs. "It is simply no longer possible," Angell says, "to believe much of the clinical research that is published, or to rely on the judgment of trusted physicians or authoritative medical guidelines." More in this case isn't necessarily better. Rosling and Angell attach different meanings to the same numbers.

What about the number of babies-per-woman whose decline Rosling finds so heartening? As usual, he has a compelling story: falling birthrates go hand in hand with economic growth, rising standards of living, and more education and employment for women. The graph tells a true story, but it's not the only story. Many economists look at graphs like this one and see a "fertility crisis." With such low birth rates, there aren't enough consumers and workers to fuel economic growth. And since deaths outnumber births in many countries, there aren't enough younger people to take care of sick, disabled, and elderly people and pay into public health insurance and retirement plans. Some sociologists suspect graphs like these may say more about economic barriers to parenting than about new opportunities for women.

Graphs and words help us. We can't do without them and neither can the numbers, because numbers can't speak for themselves. But like any good children's book, words and pictures draw us in, put us under a spell, and make us forget to question the author's story. A child who believes she can don a Superman cape and safely jump out a window will come to a tragic end. As adults making important decisions, it's just as important that we lift ourselves outside number stories, lest we get ourselves into trouble.

Some numbers are very, very big, and some are very, very small. Authors help us absorb meaning from these numbers by making analogies to things we've experienced and can imagine. How big is Facebook? By the numbers, "more than 2.2 billion people log in at least once a month." Okay, that's a lot, but with all the billions thrown around in the news every day, the number 2.2 billion doesn't give me any feeling for Facebook's power. Reporter Evan Osnos helps me out with some analogies: "If Facebook were a country, it would have the largest population on earth." Now I'm getting TV images of Chinese masses. "Facebook has as many adherents as Christianity." I have no idea how many Christian believers there are in the world, but because I'm Jewish and living in a predominantly Christian country, I have some feel for their overwhelming preponderance. And that is precisely Osnos's message: Facebook wields overwhelming power.

Climate change discussions are replete with numbers. Some seem very small, like the 1.5 degrees Celsius that is supposed to be the maximum temperature rise the earth can tolerate in the next few decades without ruination for all living things. Some climate numbers seem enormous even in their tininess, like the "400 parts per billion" that is now the concentration of carbon dioxide in the atmosphere. How about "more than 2 parts per billion," the amount by which trapped CO_2 is rising every year? These numbers make my eyes glaze. Environmentalist Bill McKibben makes them palpable with an analogy. "The extra heat we trap near the planet every day is equivalent to the heat from four hundred thousand bombs the size of the one that was dropped on Hiroshima." That image makes me shudder.

Making comparisons with things we know is probably the most important way we try to comprehend new situations. When Europeans first came to the New World, they were awed by the abundance of fish and waterfowl. In the sixteenth and seventeenth centuries, they had no way of counting wildlife precisely. They quantified with metaphors and stories. Gannets, murres, razorbills, and puffins "sat as thick as stones lie in a paved street." They compared the New World's bounty to the Old World they knew. By comparison, "the most fertile part of [all] England is (of its selfe) but barren." They found "the greatest multitude of lobsters that we ever heard of." They used their own work effort as a measuring standard. "In little more than an hour we caught with four hooks two hundred and fifty cod." (Say! What a lot of fish there are!) "In five or six hours, we had pestered our ships so with Cod fish, that we threw numbers of them ouer board againe." "He is a very bad fisher, [who] cannot kill in one day with his hooke and line, one, two, three hundred Cods." In England and France, sturgeon were so scarce as to be "king's fare," but in New England, every man "may catch what hee will."

These colorful stories about quantity don't meet today's standards of measurement, but historian Jeffrey Bolster thinks they have truth value anyway. He assesses them without imposing a twenty-first-century scientific method on earlier eras. In the sixteenth and seventeenth centuries, he notes, "metaphor was the [normal] means of conveying large magnitudes." As a historian, Bolster prizes firsthand experience and written accounts. To evaluate their credibility, he looks for consistency among different fisherman-writers, just as today's scientists look for agreement among researchers. Bolster didn't find anyone who accused another fisherman of exaggerating.

Above all, Bolster listens for the emotion between the lines. He hears the wonder, the surprise, the incredulity, the delight. "Anyone who has not seen it could scarcely believe it," a Jesuit missionary wrote from Nova Scotia in 1611. "You cannot put your hand into the water, without encountering [fish]." Exhilaration, liberation from scarcity, relief from grueling work, and the feeling of equality that comes with being able to eat like a king—these emotions all shine through the archaic language. Emotions can't be easily faked, and so Bolster thinks they authenticate the written reports. For an environmental historian, these stories provide a baseline count of fish stocks in the Atlantic. Numbers can convey quantities, but it takes a heap of well-crafted words to convey the feeling of abundance.

Sometimes a very small number shouts a very big story in the same way a Volkswagen Beetle disgorges 23 circus clowns. The magic resides in the backstories readers bring to number tales.

Friendship Park occupies land on both sides of the border between San Diego, California, and Tijuana, Mexico. Between the two countries stands a formidable fence with metal mesh just big enough for people to put their fingers through. Members of families who have been split by immigration laws and deportations gather on both sides, hoping to see, hear, and touch each other. Documentary photographer Griselda San Martin visited the park on a Sunday in 2016 when the Border Patrol opened a gate through the wall to let 5 preselected families visit in person. "Each family was allowed to embrace for 3 minutes under strict supervision from the Border Patrol and in the midst of the media frenzy," she wrote.

The event, San Martin noted, was meant to be symbolic. The U.S. Border Patrol dubbed it "Opening the Door of Hope." The Border Patrol wanted the event to symbolize the U.S. government's compassion. But only 5 families out of the thousands who've been separated by U.S. policies? And only 3 minutes? Those tiny, pitiful numbers, 3 and 5, jarred against the background stories every human heart knows—the story of irreplaceable and unquantifiable family bonds, and the story about how physical touch helps us transcend our existential aloneness. In the context of this universal emotional knowledge, those unbearably small numbers condemn the inhumanity of American border policy.

Every medical first packs a giant story into a tiny number. One of my favorite medical breakthroughs enabled a quadriplegic woman to move a robotic arm with her thoughts. The experiment involved implanting a computer chip in her brain so that she could send signals to a computer that controlled the arm. Early in the study, the doctors asked her if she had a goal. "Yeah, I have a goal," she replied. "I want to feed myself chocolate." When, after months of practice, she was able to take a nibble from a Dove chocolate bar, it was a triumphal moment. Each medical first helps only one person among the multitude who need help, but that tiny "1" symbolizes possibility and administers a powerful dose of hope for humanity. Small numbers can move us because we are symbol-making creatures. We're always on a quest to find meaning in our lives by interpreting events as instances of something greater.

We count to measure successes as well as problems. We define what we mean by success, measure those things, and if the numbers are high and growing, we tell ourselves we're successful. We keep doing things to boost our numbers. But measures of success can soon control our destiny if we don't constantly question what we mean by success.

Gross domestic product, or GDP, is the standard way to measure national economies. Politicians, policymakers, bankers, and journalists treat the GDP as a measure of societal well-being, economic development, and national productivity. GDP counts the value of all goods and services produced within a country (hence "domestic" in the name). Whether stuff is made or done by citizens or foreign nationals, if it happens inside the borders of a country, it counts toward the country's GDP.

Crucially for the meaning of GDP, goods and services are valued in money, not by, say, the number of cars a factory makes or the number of hours a nurse works. The GDP measure was designed by economists, and for them, price is the measure of value. Whatever price a buyer and seller agree upon, that's the market value. Thus, the more a drug company charges and gets for its medicines, the more value it adds to the GDP. And that is only one of many bizarre measures of success the GDP offers us.

The GDP doesn't distinguish between businesses that enhance our quality of life and ones that spoil it. Whether a factory makes antibiotics or AK-47s, its output elevates the nation's GDP. The drilling company that spent $5 billion making a colossal oil spill, throwing other companies out of business and killing marine life into the bargain—that company contributes as much to the economy as the environmental cleanup company that charges $5 billion to deal with the mess. The costs of building prison cells and security systems and of guarding and feeding prisoners all raise the GDP. If, by chance, the cost of housing, feeding, and guarding a man is higher than his salary as a teacher, he can contribute more to the nation's economic success in prison than he can as a gainfully employed community member.

The GDP doesn't care whether you produce goods or bads as long as you make money doing it. It won't judge your character, and if you want to use it to compare countries, it won't judge their character either. Like all numbers, the GDP numbers get their meaning in large part from the measure's founding story.

The creators conceived it as a measure of all things good, so everything in it is by definition good, and more is always better.

By counting only goods and services that are sold on the market, the GDP transforms our everyday sense of meaning in some perverse ways. If we believe the GDP, the sale of a $25 plastic toy contributes more to human well-being than an hour of home care provided by an aide who is paid $15 per hour. According to the GDP, a mother who buys Alexa to entertain her child is a better parent than the one who checks out library books and reads them to her child. The GDP thinks a million-dollar prime-time ad for beer enhances the nation's well-being three times over: first, by helping a TV station produce more programming; second, by helping breweries sell more beer; and third by helping rehab facilities sell more detox services. And as far as the GDP is concerned, no thanks are due the volunteers who rush to disasters, staff food pantries, tutor kids, mentor immigrants, and keep nonprofit organizations running. The GDP ignores their contributions because they don't move money around.

The GDP is a measure of total production—the bottom line, so to speak. It pays no attention to how all the riches are distributed. Of course, the residents of each country do care about "how much of it is mine," especially if they have very little. To compare the standard of living in different places, economists calculate each country's per capita (or per person) GDP. Per capita GDP is a fantasy measure if ever there was one, because it pretends that a country's total wealth is divided equally among its population. The people at the bottom know better. They also know that money isn't their only source of sustenance. They may have only $2.00 a day to live on, but they usually partake of a cashless economy, rich with bartering, task sharing, and everyday help. The GDP includes only what gets bought and sold in markets. It misses everything else that contributes to people's well-being—strong social connections, nature, making art and music for the sheer pleasure, social harmony and peace, and time to enjoy all of the above.

Every number gets some of its meaning from dramatic packaging and the GDP is no exception. In the U.S., GDP is announced with theatrics worthy of a Broadway blockbuster. Toward the end of each quarter, government accountants enter a "lockup" in the Commerce Department, where they run their numbers in secrecy. When they're finished, they put the results in a sealed envelope that they hand off to guarded messengers, who take it to the White House. There, the chair of the president's Council of Economic Advisers checks the numbers and approves them before they're delivered to the president. All this time, the GDP remains top secret. The next morning, the GDP is announced to reporters, but not before they're locked in an electronically sealed conference room to prevent them from emailing ahead of the big public revelation. Inside the secure chamber, reporters receive the new economic report and have one hour to draft their stories. Instead of raising a curtain, officials flick a switch so that reporters can electronically deliver their lines to their audiences.

No matter what the GDP number turns out to be, the drama seduces the public into believing that this number embodies our national wealth. Just as theater audiences throw themselves into a play by turning off their reality checkers, politicians, bankers, reporters, and everyone who follows economic news all play along with the fiction. They accept it not because they lack critical thinking, but because, well, everyone else in the theater is having a good time by pretending that what's on stage is real. The GDP isn't a valid measure of well-being or standard of living, but like most scientific measures, it draws a good deal of its support from dramatic techniques.

Critiques of GDP aren't exactly news. Even the economists who preach and teach about GDP acknowledge its faults. They keep using GDP and other distorted economic measures because, they say, some measure is better than none. Without our "indispensable" economic tools, warned a former chair of the Council of Economic Advisers, "we would be in the economic dark ages." Paul Samuelson and William Nordhaus, two famous economics textbook authors, don't send us all the way back to the Dark Ages, but they're not above scare tactics either: "Without measures of economic aggregates like GDP, policymakers would be adrift in a sea of unorganized data."

Before we jump aboard the GDP ship, we ought to ask ourselves where the captain is headed and whether we want to go there. The GDP has been taking us to places where oil spills are prized for the jobs they create and work done out of the goodness of our hearts counts for nothing. As one environmental economist put it, "We tend to get what we measure so we should measure what we want."

In the sea of unorganized data, the GDP ship still dominates and carries the big guns, but alternative approaches to measuring societal well-being are gaining steam. Other measures attempt to count everything from health of the environment to human development to happiness. One alternative, the Genuine Progress Indicator, works by adding good things into the GDP, such as the value of highways and volunteer work, and subtracting bad things, such as the costs of crime and ozone depletion. Canada and several U.S. states use this indicator in their policy planning. The European Union has a "Beyond GDP" commission with a mandate to explore alternative ways of measuring growth and progress. But for sheer delight, Bhutan tops them all with its famous Gross National Happiness measure.

"Closing averages on the human scene were mixed today. Brotherly love was down two points, while enlightened self-interest gained a half. Vanity showed no movement, and guarded optimism slipped a point in sluggish trading. Over all, the status quo remained unchanged."

Alas, I can't take you to Bhutan, but I can take you on a tour of the Social Progress Index made in Washington, DC. The index is laden with interesting ideas about quality of life as well as thought-provoking measurement problems. It is not the last word, nor do its authors think it is, but it represents a welcome departure from the GDP. Instead of starting with market transactions and adding or subtracting goods and bads, the Social Progress Index includes 51 measures of health, education, security, environmental quality, political equality, and freedom. (I bet there's a story behind how the fifty-first measure made the cut.)

The general thrust of the Social Progress Index seems to go in the right direction—which is to say candidly what scientists aren't supposed to admit: I like its values. For example, to measure personal freedom, it includes not only the usual freedoms of religion, expression, and political participation, but also access to modern family-planning methods, acceptance of ethnic minorities and gays, and girls' freedom from childhood marriage. For some people, those freedoms mean social decay, not progress. Others might wish for different items to be included in the basket of indicators. Perhaps arrest and incarceration rates should be used to measure lack of personal security or freedom. Why not ease of starting a new business?

Some loyalists on the GDP ship fault the new measures for being ideological and emphasizing what their creators value. But that's exactly why the renegades jumped ship—to count what they value that the GDP ignores. Hewing to old, established counting rules like the ones in the GDP enable people to watch trends over time, but at the cost of stifling debate about values. Even though no measure of something so grandiose as social progress can reflect everyone's values, the exercise of trying to design a measure is valuable in itself. Merely thinking about what's important to measure forces us to clarify our values. And if we remember that every counting rule is a human judgment, ours to make and ours to question, then counting can be an opportunity to let our imaginations roam with our values.

So far my literature lessons have lacked a good whodunit. Enter the Flint, Michigan, water crisis.

Flint was a thriving manufacturing city until General Motors decided to close its auto plants there. Flint became known as "the unemployment capital of the U.S." The city survived with the help of federal, state, and foundation money, but in 2014, Michigan's governor deemed the city incapable of managing its finances. Using a state emergency powers law, he suspended the local government and appointed an emergency manager to bring efficiency to the city. As one of his cost-saving strategies, the emergency manager switched the city's water supply from distant Lake Huron to the nearby Flint River.

Soon after the switch, residents noticed that the water coming out of their taps looked brown and murky, smelled foul, and tasted bad. They asked to have their water tested. Within a few weeks, people learned that the city water was contaminated with lead. After enough citizen complaints, the Michigan Department of Environmental Quality tested the water. It wouldn't release its numbers but announced many times that the water was safe to drink. "Let me start here," the department's spokesperson began a radio interview. "Anyone who is concerned about lead in the drinking water in Flint can relax."

Flint citizens couldn't relax because two scientific agencies had given them cause for concern. Lead has long been known to cause brain damage in both adults and children, but it's much more harmful to young children and to fetuses in utero. Lead can stunt children's development, lower their IQs, and cause lifelong intellectual, physical, and behavioral problems. Knowledge about these dangers prompted bans on leaded gasoline and lead paint and new regulations to control lead in water. Health experts believe there is no safe level of lead in drinking water, but the EPA (Environmental Protection Agency), in its Humpty Dumpty-ish wisdom, says that only levels above 15 parts per billion require steps to reduce the amount of lead in water. (Never mind "parts per billion." Just remember the number 15.) The EPA's standards signal what lead-in-water numbers mean. Any number above 15 means you should worry.

Meanwhile, the CDC (Centers for Disease Control and Prevention), the country's top public health agency, sets standards for dangerous levels of lead in blood. Before 2012, the CDC thought 10 (micrograms of lead per deciliter of blood) was worrisome and called 10 "the level of concern." In 2012, the CDC lowered the level of concern to 5 and clothed the number with a new name that conceals anything provocative. Now it calls 5 a "reference level," a name even more opaque than the milquetoast "level of concern." And in true Humpty Dumpty fashion, the CDC recommends that no medical treatment should be given unless a child's level reaches 45. The CDC's names and recommendations signal what the numbers should mean to parents and doctors. But the gap between 5 (concern) and 45 (treatment) leaves a lot of room for confusion—a room where parents' worries and officials' complacency can both grow.

In Flint, only a few citizens knew the numbers, but most knew enough about lead poisoning to worry. They didn't believe the city's assurances. They believed their eyes, noses, and taste buds, their skin rashes, their clumps of hair falling out, and their children's illnesses. They held their own "trial" and brought in two sets of numbers to testify about what was going on. Numbers being only numbers, they had to be accompanied by the people who had created them.

First witness: *Professor Marc Edwards, a Virginia Tech expert on water contamination.*

A citizens' coalition invited Professor Edwards to test their household water. He gave out water-testing kits and explained how to collect samples. Professor Edwards and his citizen team found lead levels of almost 28 in over 90 percent of the homes it tested. Some homes had levels high enough to qualify as "toxic waste" (really bad) or "hazardous waste" (really, really bad) by EPA standards. Edwards also identified three neighborhoods within Flint where the water's lead levels were the highest. All three are predominantly black.

Second witness: *Dr. Mona Hanna-Attisha, a pediatrician and public health researcher at Hurley Medical Center.*

Hurley Medical Center is the place where all Flint children under 5 have their blood tested for lead—if they get tested. Shortly after the city switched its water source, Dr. Hanna-Attisha began hearing concerns about the water from her patients' moms. She was aware of the news stories, citizen protests, and Professor Edwards's water tests. She decided to compare children's blood tests before and after the water switch. Sure enough, she found many Flint children whose blood lead levels had risen above 5 after the switch. Before the switch, 2.4 percent of Flint kids had elevated lead levels; after the switch, 4.9 percent did. In the three wards with the highest water-lead levels, 10 percent of children had blood levels above 5 after the switch.

Compared with the EPA's water standards and the CDC's blood standards, both sets of numbers were way too high. The numbers served as compelling witnesses to something bad happening in Flint, but by themselves, they couldn't identify the perp. What did the numbers say about responsibility? Who or what made the numbers high?

Professor Edwards and Dr. Hanna-Attisha had their strong suspicions. Edwards is a civil engineer. He immediately suspected the water pipes, because lead pipes corrode easily and leach lead into the water. Lead pipes were standard up until 1986, when the Safe Drinking Water Act banned them for new construction. Flint didn't have a lot of money to upgrade its city pipes, nor did most residents, especially those in poorer neighborhoods. The EPA requires cities to treat their water with substances to control corrosion. Edwards guessed the city wasn't treating the Flint River water. His numbers, combined with his knowledge of how lead gets into water, steered him to some human perps. The numbers wouldn't and shouldn't have been so high if the city had used corrosion control after it switched water sources. Dr. Hanna-Attisha's before-and-after blood-level results clinched the case for him.

Dr. Hanna-Attisha came to her suspicions by another route. She was tipped off by a city health department employee who let slip that the department had been testing children and had noticed a "spike" in lead levels a few months after the switch. Hanna-Attisha asked many times to see the test results. The employee stonewalled and never sent them. Dr. Hanna-Attisha had also learned from Professor Edwards that the city inspectors were using faulty procedures to test water samples. The inspectors let tap water run for 25 or 30 minutes before taking their samples—plenty of time to flush most particles out of the pipes. Hanna-Attisha put 2 + 2 together: City leaders were deliberately minimizing the water numbers and hiding the blood test numbers. Many people doubted that leaders would knowingly poison their citizens, but not Hanna-Attisha. Her parents are Iraqi and had lived under Saddam Hussein. She had seen photos of dead babies and adults killed by Saddam's poison gas. She knew full well that political leaders could poison their people.

The city and its citizens fought over the meaning of the numbers by clothing them with words. The city and its defenders claimed no one was responsible because there was nothing to be responsible *for*. Lead occurs naturally in water. Flint River water had different chemicals and minerals than the water the city had been getting from the Detroit water authorities. The different smell and taste was "natural," something the citizens would eventually get used to. One city defender made sure to absolve the Flint River itself: "The issue with the [lead in Flint's] drinking water stemmed from the Flint River having natural differences in chemistry compared to water from Lake Huron. The river water itself was not to blame." But neither were any people to blame, according to the city's leaders.

The term "lead poisoning" was pretty standard in public health circles. "Poison" suggests an evildoer. Dr. Hanna-Attisha used the word "poisoned" liberally and blamed the city for the children's high lead levels. So did the citizens and the media. Some doctors at Hurley Medical Center, the same hospital where Dr. Hanna-Attisha works, insisted that children weren't "poisoned" by the city, they were simply "exposed" to lead. To call them "poisoned" assumes they're "irreparably brain damaged" and "unjustly stigmatizes their generation." A group of Hurley physicians even voted to banish the words "lead poisoning" to describe the Flint situation.

Soon after Dr. Hanna-Attisha announced her blood test results, the state governor acknowledged the city's failure to keep Flint citizens safe. Under duress, the state and city released documents and emails they had kept secret. Michigan Environmental Quality officials had done everything possible—against EPA regulations—to make sure water tests didn't show high levels. And when the tests did come back high despite their efforts, they lied about the results. Health officials had concealed high blood-lead tests that should have triggered public health actions to monitor children and fix the lead contamination. Within three years of the water crisis, the top officials in Flint were indicted for multiple crimes.

More often than not, as we'll see in the next chapter, numbers are weapons of the powerful. The Flint story shows how the weak can wield numbers against the strong. In the Flint water crisis, numbers didn't identify the perps—people did—but numbers were the crucial witnesses to the crime. The EPA and CDC standards gave the water and blood numbers meaning. Professor Edwards and Dr. Hanna-Attisha used the numbers, replete with meaning, to prove the existence of a problem. Numbers in hand, they went looking for human causes. As in most successful resistance movements, the Flint victims had help from some powerful allies—not only Professor Edwards and Dr. Hanna-Attisha, but also other experts and journalists willing to investigate the story and give it national play. Nevertheless, when all was said and done, numbers were the silent heroes of Flint.

4

How Numbers Get Their Clout

I'd lived in Boston for years and knew when to expect rush-hour snarls in different parts of the city. Then one fall, after I returned from summer vacation, no matter where or when I drove, the traffic seemed much heavier than I remembered. I couldn't rely on my traffic savvy anymore, so I started allowing more time for trips. Yet even though I trusted my sensations enough to act on them, I wondered whether the increased traffic was a measure more of my impatience than of the number of vehicles on the road.

The following March, the *Boston Globe* ran a front-page story headlined "Boston's Clogged Arteries." The article compared old bus schedules with current ones and calculated how much longer it took buses to get in and out of the city compared with 10 years earlier. It cited studies that said Boston drivers were spending 2 hours more in traffic in 2017 than they did the year before. A cute graphic displayed little stopwatch images and lots of numbers.

INCREASING DELAYS

60

Hours spent in traffic for Boston drivers in 2017, a two-hour increase from 2016

More minutes for DATTCO bus's 6:50 a.m. ride from Fairhaven to Back Bay, compared to 10 years ago

More minutes for Boston Express bus from North Londonderry, N.H., to South Station since 2008

Longer for MBTA morning express routes from Brighton, Watertown, and Waltham, compared to 2017

"Phew!" I thought. "I'm not getting prematurely grumpy. It's not in my head. There are numbers to prove it." My relief didn't last long, though, because I was writing a book to knock numbers off their pedestal and here I was, trusting numbers more than my own experience.

What is it about numbers that makes us put so much faith in them and trust them as oracles of truth? After all, when numbers speak, they summarize judgments that humans have already made about "what counts." Numbers acquire their power the same way the gods acquire theirs—humans invest them with virtues they want their rulers to have. Call it fairness, call it lack of favoritism, call it meritocracy, call it equal treatment. Call it wisdom. Or, as many people do now, call it objectivity. Our numbers, like our gods, promise to govern us well.

People didn't always trust numbers as readily as we moderns do. Before the eighteenth century, measuring sticks and containers weren't standardized as they are now. Counting was a local and customary affair. In every relationship—lords and serfs, landlords and tenants, merchants and farmers—the powerful controlled the instruments of counting and used them to practice what we now call portion control. They counted "low" to extract more produce, labor, and taxes from their subjects. Local tax collectors took payments in kind. In one French town, the miller measured dried corn kernels when he was collecting from the farmers, but swelled the kernels with water before measuring out the portion he owed to the lord. He kept the surplus for himself. A miller might pay a farmer for his grains by heaping the measuring container so high that no more particles would stay. Then when it came time to measure out grain owed to the lord or seigneur, he'd be sure to level off the container. Knowing the gains to be had from heaping measures, the seigneur and tax collector could redesign a measuring container with a broader base so it would hold a higher heap. Their tactics were obvious, but their measuring methods were law.

Thus, for the people on the weak end of these relationships, measuring tools and the numbers they yielded were hardly vessels of truth. They were opportunities for bosses and merchants to stint, cheat, and weasel out of a fair bargain.

By the end of the eighteenth century, support for a standardized measuring system was easy to come by. While French peasants were slowly fermenting their rise up to revolution, French scientists were cooking up the metric system. Their recipe pegged a standard unit of length—the meter—to the distance between two points on the earth's surface. It sounded like a good idea at the time. What could be more fixed and more natural than the ground we stand on? The surface of the earth was immune to human influence and couldn't be altered by pecuniary motives, or so they believed.

The Academy of Sciences hired two engineers to survey the length of a meridian from a town in northern France to Barcelona. (A meridian is one of those imaginary curved lines running around the globe from the North Pole to the South Pole.) Each man scouted out the highest hills, steeples, and towers on which to perch his survey instruments. Sighting from one point to the next, each measured short distances until they met in the middle. Add 3 scoops of geometry, 2 scoops of astronomy, a dollop of imagination, *et voilà*! A meter is exactly 1 ten-millionth of the distance from the North Pole to the Equator.

Never mind that the earth and its people didn't always make the job easy. The surveyors couldn't get permission to use some of the tallest buildings. Some of the structures they planned to use no longer existed. Never mind, either, that the two men had to make seat-of-the-pants adjustments to their finely calibrated measurements. One church steeple leaned to the west, throwing off the measurement. Some of the towers were so badly decayed that the scopes couldn't be leveled. On some days, freezing temperatures hurried the men into making less-than-careful observations. Never mind all that serendipity. The scientists did their best and in 1792 the meter was given physical form in a pure platinum stick. The platinum stick, in turn, was given its first rush of political authority when it was presented with great fanfare to the new, democratically elected French legislature.

A length of a merchant's cloth or the area of a farmer's field would no longer vary from village to village, no longer depend on measuring rods owned and created by the local powers that be. From then on, there would be "one weight, one measure," as the Revolutionary slogan proclaimed. The people's craving for deliverance from local tyrants was strong enough to invest the scientists' concoction with almost holy truth. The meter was the next best thing to a piece of the earth itself. It made every landowner "a co-owner of the World," waxed one of its proponents. Iron replicas of the platinum meter stick were presented to foreign scientists. A Dutch astronomer felt his iron stick beginning to forge "fraternal bonds" among European peoples.

That's a lot of magical power for a metal stick fashioned by human hands and minds. These eighteenth-century testimonials to the meter stick sound overblown to our ears, but the democratic promise of standardized measures hasn't lost any of its allure. In the mid-twentieth century, Witold Kula ended his magisterial history of measurement with an ode to the metric system: "Gone are the countless daily opportunities for the strong to injure the weak, for the smart to cheat the simple, and for the rich to take advantage of the poor." Would that it were so.

The metric system seemed to protect weak villagers from their local heavies, but no measurement system enforces itself. Standards of any kind—counting methods, legal rules, or quality criteria—must be defined and enforced by someone. Just as nobles had once imposed their preferred measures on the people they ruled, France's new national government imposed the metric system on the nation.

Even though the new government was obviously throwing its weight around, French citizens had fought for a centralized government and they convinced themselves that democracy gave them more control over their local nobles. No doubt the "little people" experienced a greater sense of fairness when they saw that the same counting methods applied to everyone. A uniform counting system, people believed, prevented officials from manipulating the facts. Decisions based on numbers came to seem more objective than personal judgments. Uniform measuring systems became associated with equality and fairness. A platinum stick, after all, couldn't play favorites. It couldn't count low for the poor and high for the rich.

Rebelling against tyranny is easy compared to what comes next. With no common enemy to fight, people begin to fight with each other. Who's going to referee the fights? Over in the New World, the American Founding Fathers put their faith in a different set of numbers. The U.S. Constitution begins by setting forth the structure of Congress—in other words, how power will be granted to the people. Here's a not-quite-verbatim summary, replacing number words with numerals so they stand out:

> *Representatives to the House are to be chosen every second year by the several states. To be chosen as a Representative, a person must have attained the age of* **25**, *and must have been a citizen for* **7** *years. Representatives and taxes shall be apportioned among the states according to their respective* **numbers**. *An* **enumeration** *must be made within* **3** *years after Congress holds its first meeting, and every* **10** *years thereafter. The number of state representatives shall not exceed* **1** *for every* **30,000** *but each state shall have* **1** *Representative. Until the first census is complete, New Hampshire shall have* **3**, *Massachusetts* **8**, *Rhode Island and Providence Plantations* **1**, *Connecticut* **5**, *New York* **6**, *New Jersey* **4**, *Pennsylvania* **8**, *Delaware* **1**, *Maryland* **6**, *Virginia* **10**, *North Carolina* **5**, *South Carolina* **5**, *and Georgia* **3**. *Each state shall have* **2** *senators, chosen for* **6** *years, and each will have* **1** *vote. Each senator must be at least* **30** *years old and must have been a citizen for at least* **9** *years.*

Article I reads like instructions for assembling a precision machine, except that its authors slipped in an explosive device. The states pay taxes and get congressional representatives according to their "numbers." Some unspecified agent shall carry out an "enumeration"—of, um, what? In my summary, I left out the only clause that uses the word "persons," the one that explains how each state's population—its "respective numbers"—would be determined.

How to count the population was the most contentious issue at the Constitutional Convention. The size of each state's population would determine how much it had to cough up in taxes for the federal government, as well as how many seats it would hold in the House of Representatives. If slaves were counted as items of property, their owners would be taxed on them; if slaves were counted as people, the Southern states where almost all of them lived would get more representatives in Congress. (You see, we're back to the classification problem.) In the infamous compromise ultimately written into the Constitution, slaves were counted as three-fifths of a person.

Three of the Founding Fathers—Alexander Hamilton, James Madison, and John Jay—published essays about the sticking points of the draft constitution in order to gather political support and persuade the states to ratify it. These essays became known as *The Federalist Papers* and they were numbered in the order in which they were published. In Number 54, James Madison defended three-fifths as the correct value of a slave.

Madison was a slaveholder. He had white skin in the game, not to mention that he saw himself as defending the South's political economy. Instead of acknowledging his partisanship, though, he created a fictitious Southerner and let him argue against the northern position that slaves should count as property, not people.

Madison has his imaginary Southerner work through a phony legal analysis to prove that slaves "are considered by our laws in some respects as persons and in other respects as property." Slaves, the Southerner says, have 3 legal characteristics that make them appear as property. First, a slave is "compelled to labor not for himself, but for a master." Second, a slave is "vendible [can be sold] by one master to another." Third, a slave is "subject at all times to be restrained in his liberty and chastised in his body." In all these ways, the slave "may appear to be degraded from the human rank and classed with those irrational animals which fall under the legal denomination of property."

So far the North seems to be winning. If slaves are property, Southerners will pay more in taxes. Madison's imaginary Southerner quickly mounts his offense. There are 2 ways the law considers slaves as people, he says. First, a slave is "protected, in his life and limb, against the violence of all others," including his master. (That must have been news to both slaves and masters.) Second, a slave "is punishable himself for all violence committed against others." In these ways, the law regards the slave "as a member of society, not an irrational creation."

From this feat of sophistry, Madison's alter ego concludes: "The Federal Constitution therefore, decides with great propriety on the case of our slaves, when it views them in the mixt character of persons and property. This is in fact their true character." Madison then returns to his own voice, pretending to be an unbiased judge of the debate, as if his readers don't know he's a Virginian and a slaveholder. Speaking as a neutral judge, toting up the weight of evidence on both sides, he allows as how the imaginary Southerner's reasoning seems "a little strained," but "on the whole" it persuades him that the Constitution establishes the correct "scale of representation."

I'm not sure of Madison's math skills. By my count—3 points for slaves as property, 2 points for slaves as people—each slave should have counted as only two-fifths of a person. But math aside, I don't know a more searing example of how counting rests on metaphorical reasoning: a slave is like property if we look at certain traits and like a person if we look at others. *Federalist* Number 54 sounds like the kind of "compare and contrast" essay you might have had to write for a high school composition class. But Madison wasn't writing to show off his lit-crit skills. He was writing to preserve the economic and political power of the South. For that, he needed a killer metaphor. He equated the law's treatment of slaves with their essential traits, their "true character."

To find evidence about the important attributes of slaves, Madison didn't consult ministers or philosophers. Neither did he ask his slaves. He looked to laws. In other words, he consulted political decisions that had already been made by dominant white elites. From those acts of raw power, he purported to find the standard of truth for how government should make further political decisions. That's a lot of magical power for a fraction fashioned by human hands and minds.

Two hundred and some years later, we're still using numbers to preserve the laws of slavery, only now we're doing it with computers and algorithms instead of quills and parchment. About half the states and hundreds of cities use algorithms to decide people's fate in the criminal justice system. Algorithms recommend whether to release suspects on bail, or sometimes with no bail, pending their trials. Algorithms suggest appropriate sentences for convicted criminals. They also advise which prisoners are likely to go on to commit more crimes if they're released on parole. All these recommendations come in the form of numerical risk scores the algorithms assign to defendants. This one's a 3, low risk, good candidate for parole; that one's a 9, trouble on wheels, best to keep the doors locked.

Julia Dressel was a computer science and gender studies major at Dartmouth College when she decided to investigate the most widely used algorithm in criminal justice to see just how accurate and fair it is. COMPAS, the algorithm, had recently been in the news after ProPublica, a public-interest research group, released a study claiming that it was highly inaccurate in predicting which parolees would commit future crimes. ProPublica's researchers had a reality test: they had data on actual defendants for two years after the algorithm had made its predictions. They knew which parolees committed crimes and which ones didn't. COMPAS, they found, got it wrong 35 percent of the time. Moreover, COMPAS erred in ways that were much tougher on blacks than whites.

Of the black defendants the algorithm deemed *highly likely* to commit another crime within two years, nearly 45 percent did not. By contrast, only 23.5 percent of white defendants were wrongly labeled as likely future criminals. Almost half (about 48 percent) of the white defendants who did commit another crime were "let off" by the algorithm as unlikely to reoffend, but only 28 percent of black defendants who actually committed another crime were given a pass by the algorithm. These numbers, ProPublica said, showed that the COMPAS algorithm was racially biased.

The rationale for using algorithms in criminal justice decisions is that they are high-tech platinum sticks. They can't harbor prejudice and they can't be swayed by emotion or self-interest. The criminal justice system has been infected with racial bias since Madison's time, so any method that promises racial equality and any other kind of equal justice is to be welcomed.

After reading the ProPublica study, Julia Dressel wondered whether the COMPAS algorithm was really capable of judging more accurately and fairly than humans. She asked 400 people who had no criminal justice experience to judge the same defendants as the ProPublica study of COMPAS had used. Dressel recruited participants through Amazon's Mechanical Turk, a kind of online temp-work platform where anyone can sign up to answer survey questions and get paid a small fee. The COMPAS algorithm had 137 pieces of data on each defendant to work with. Dressel gave her participants only 7 pieces of data about each defendant—sex, age, and 5 items about previous convictions. She asked her judges-for-a-day only one question, the same one COMPAS answered: "Do you think this person will commit another crime within two years?"

To Dressel's dismay, the people off the street with no criminal justice experience performed just about the same as COMPAS. Overall, they judged correctly 67 percent of the time, compared with the algorithm's 65 percent. The human judges wrongly predicted that 37 percent of innocent blacks would reoffend, slightly less than COMPAS's 40 percent rate. The humans "let off" 40 percent of the whites who actually committed crimes within two years, quite a bit lower than COMPAS's get-out-of-jail-free cards for almost 48 percent of the white reoffenders.

What was going on inside the algorithm to produce such poor predictions and such racially biased ones? Dressel figured that if her untrained human judges with only 7 pieces of data could do just as well as COMPAS with its 137 information bits, maybe the algorithm had a lot of built-in excess. She used her math skills to write a simpler algorithm using only the 7 data bits her humans had used. (For you statistics mavens, it was a logistic regression model.) The stripped-down model did just as well as COMPAS. Dressel wondered whether she could shed even more data bits and still get similar results. She built a new model with only 2 pieces of information, the defendant's age and number of previous convictions. Her 2-bit classifier performed just as well as COMPAS.

If 2 pieces of data are just as good as 137, we might well ask what purpose those extra 135 data bits serve. Are they a screen like the one Toto knocked over in the Land of Oz, exposing a frail old man pretending to be a wizard? Are those 135 superfluous data bits just so much razzmatazz that persuades politicians and judges to let algorithms take over their criminal justice systems? Cities and states pay big bucks to private companies for COMPAS and other similar algorithms. Are those 135 extra data bits meant to make a platinum stick look like a Rolls Royce and sell for as much?

Leaving aside the 135 apparently useless data bits, there's something troubling about the 2 information bits that seem to do all the heavy lifting for the COMPAS algorithm. One of them is the defendant's age. Younger defendants get higher risk scores. Of course, the younger people are, the longer they can expect to live and the more time they will have to commit crimes. Or, for that matter, the more time they might have to do good. But the algorithm isn't predicting a lifetime. It's only predicting behavior in the next two years after the defendants' release. By consigning younger people to remain in prison while they await trial or another shot at parole, the algorithm separates a cohort from education and formative social experiences that could enable it to become another Greatest Generation.

In Dressel's super-simple model, the second predictive data bit is the number of a defendant's past convictions. If age and criminal history are all it takes to predict the risk of committing a future crime as accurately as COMPAS does, COMPAS is apparently betting that someone who's been a bad actor before is going to keep on acting badly. Indeed, many studies have found that criminal history *is* the best predictor of whether someone will skip bail or commit another crime, so it's not surprising that algorithms would give criminal history an outsize influence on defendants' scores. In fact, another widely used risk-prediction algorithm, called Public Safety Assessment, uses only age and criminal history in its database, quite like Dressel's 2-bit classifier.

But there's another way to read the mind of an algorithm that relies so heavily on criminal history. Maybe it's not even looking at defendants. Maybe it's looking at police and judges and betting they're going to keep on arresting and convicting the same people they've busted in the past. That's a smart bet, because the criminal justice system does tend to keep going after the same people.

It's hard to quarrel with the motivations behind predictive algorithms. The people who design and use them have the best of intentions. They understand the shortcomings of the criminal justice system and hope mathematical tools will improve decision making. They want to keep the public safe by not releasing dangerous criminals. They want to stop locking up people who pose no threat to society. They want the system to stop treating black and brown people more harshly than white people. Surely, they believe, algorithms can do all these things because they're more accurate and objective than humans.

Advocates of algorithms believe that judges, prosecutors, and hearing officers tend to go by their intuitions and the rules of thumb they've developed over the years. They're subjective. They decide on the basis of their "instincts and experience." Sometimes their personal biases sway them, albeit unconsciously. And too often, they're terribly wrong. Algorithms help these fallible people make evidence-based decisions. Algorithms "can give judges an objective and scientific measure of risk" because they incorporate massive amounts of data in their calculations. Every data bit (so the argument goes) is a piece of factual evidence that can help determine the truth.

Before we accept this argument, we ought to look closely at where the data bits come from. We tend to think of criminals as bad people and crimes as things that bad people do, but it takes a village to make a crime. Contraception, interracial marriage, homosexual relationships, and selling alcohol have all been crimes at some point in U.S. history and no longer are. Why not? Not because the former criminals changed their behavior, but because legislatures and courts changed theirs. Legislatures and courts define the rules that place an activity on the right or wrong side of the law. Police and other enforcers look for activities on the wrong side—that's their job—and in the process, they make snap judgments about suspicious behavior and whether to arrest. Prosecutors decide whether to bring charges and what kind. Opposing lawyers choose which evidence to present. Judges and juries weigh the evidence and come to a decision—crime or no crime, guilty or innocent.

Predictive algorithms suction up all that government power, call it data, and bottle it in mathematical formulas. The data are factual in the sense that they record things that actually happened. But think about what actually happened. Officers in the criminal justice system made stops, arrests, charges, convictions, and sentences. That means the data incorporate the officers' intuitions, instincts, rules of thumb, and biases as they decided how to enforce the law. The subjectivity that algorithms supposedly avoid resides deep in their data sets, only now it's much harder to see.

And here is where math meets Madison. Madison examined how the law treated slaves—as salable property without free will. Then he converted those utterly human decisions into inborn character traits. Since those traits are how slaves *are*, Madison insisted, it is correct to count them as part human, part property. Slaves became three-fifths of a person. Three-fifths of a person remained a slave.

In today's criminal justice system, predictive algorithms examine how the law treats people. The algorithms give great weight to arrests, convictions, and sentences. Then they convert those utterly human decisions into character traits of the people the system has already swept up, only now character traits go by the more scientific-sounding name of "risk." Defendants become six-tenths of a criminal or whatever risk score an algorithm assigns. Sixth-tenths of a criminal remains in jail.

If we don't want to maintain the laws of slavery, metaphorically speaking, it behooves us to delve into how the categories inside our counting practices were established and what power relationships defined them. If we simply rely on categories and counting rules embedded in past practices, we're perpetuating the same power relationships that wrote the laws in the first place.

Child protection agencies work under a terrible moral burden. Predicting whether a child is at risk of abuse or neglect is a judgment call and the consequences of guessing wrong can be devastating. Failing to predict abuse means a child will be severely harmed, possibly killed. But falsely predicting abuse will stigmatize a family and could lead to removing a child unnecessarily. With such heavy responsibility, child welfare specialists yearn for a reliable way to make sure no child whose case they handle becomes a victim of severe abuse—and a headline. "That's why I love predictive analytics," one prominent child abuse researcher told a *New York Times* reporter. "It's finally bringing some objectivity and science to decisions that can be so unbelievably life changing."

The Department of Human Services in Allegheny County, Pennsylvania, employs the usual panoply of highly trained social workers. Since 2016, it has also employed an algorithm to help identify children at risk of abuse or maltreatment. A "Data Warehouse" provides the human staff with information from about 30 programs that low-income people use, including medical care, housing, day care, and food assistance, and from school, police, jail, and court records. In addition, the staff relies on hotline calls from anyone in the community who has concerns for a child's welfare, and from professionals, such as teachers, day-care providers, and doctors, who are legally mandated to report their suspicions of abuse, even if their suspicions are based only on hearsay.

Every time a call comes in to the department, human screeners check the allegations in the call against data in the Warehouse to decide whether the situation sounds serious enough to investigate further. With data in hand, the screener makes an educated guess about the likelihood and severity of possible abuse. Screeners use simple categories, not numbers: risk can be low, moderate, or high; danger can be "no safety threat," "impending danger," or "present danger." Then the screener presses a blue button to find out what the algorithm thinks. The algorithm displays its judgment as a number from 1 to 20, along with an image reminiscent of a thermometer or a fire warning. High scores are displayed at the top in a red-orange alarm zone. Low scores are shown at the bottom in a cool green meadow.

How does human judgment compare with the algorithm's? Political scientist Virginia Eubanks sat with an intake screener one day to match wits with the algorithm. Together they considered two families who had been reported to the agency. One case was a 6-year-old boy whose mom had gone to a therapist for help with anxiety. Mom told the therapist she had found her son outside on the porch crying and was concerned that something bad might have happened to him. The therapist reported the family to the agency. A week later another call came in from a homeless services agency reporting that the boy had dirty clothes and poor hygiene, and there were "rumors" that the mom was doing drugs.

The other case was a 14-year-old boy whose family had received a host of social services. When their case manager visited on a November day, she found a cold house with a broken window and door, a cluttered living room, and a smell of pet urine. The teen was wearing several layers of clothes.

Eubanks thought neither youth was in any serious danger. Leaving a kid on a porch isn't neglect, let alone child abuse. Neither is having inadequate heat or less-than-pristine housekeeping and grooming. As Eubanks points out, the signs of risk pinned on these two families are routine features of poverty. She guessed the 6-year-old should get a score of about a 4 and the teen a score of 6. When the algorithm spoke, it scored the 6-year-old at 5 but the teen got a 14. In practice, the screeners don't assign numerical scores, as Eubanks did. She tried her hand at scoring to see whether she could get insight into how the algorithm works and into what happens when the human and algorithm disagree.

When human and machine judgments are far apart, a delicate dance ensues. In theory, the algorithm is a tool to help intake screeners do their jobs. The text accompanying the thermometer image says the tool is "only intended to help inform call screening decisions" and "[shouldn't] be considered a substitute for clinical judgment." The algorithm's score doesn't dictate whether the department should swoop in and remove a child. A high score suggests that the department should "screen the case in" for further investigation. In most cases, the decision whether to investigate is up to the screener, but high scores and certain other criteria put a case into the category of "mandatory investigation." Even there, screening supervisors can override the tool's judgment.

So far in the tool's three-year career in the department, managers and frontline staff seem to have different views about trusting it. Screeners tend to rely on their own scores more than the algorithm's score and screening supervisors override 1 out of every 4 mandatory investigations. In an evaluation commissioned by the department, only half of the 16 screeners interviewed said they had confidence in the tool. Several months later, after the tool had been enhanced, 13 out of 18 screeners said the algorithm "occasionally" generated a score they felt was inaccurate and another 2 said they "frequently" encountered inaccurate scores. What do screeners do when they think the algorithm's score is wrong? Half said they go to their supervisor. A few said they recheck the department's data file and review their own research. Only 1 person admitted to relying on research instead of the algorithm. Some screeners felt that "machines can't take the human element into account" and "can't recognize information that needs to be updated."

By contrast, most managers think the machine has better judgment than humans. One manager told Eubanks, "If you get a report [a phone call] and you do all the research, and then you run the score and your research doesn't match the score, typically there's something you're missing." In other words, the machine is usually right. Erin Dalton, leader of the department's data analysis section, is somewhat harsher on the screeners. "They want to focus on the immediate allegation, not the child's future risk a year or two down the line. They call it clinical decision-making. I call it someone's opinion." Dalton hopes to get the screeners to "trust that a score on a computer screen is telling them something real," but it's difficult to change their mindset. "It's a very strong, dug-in culture." Someone on Dalton's staff elaborated on her view in an email to me. Dalton worries that the screeners might give more weight to one hotline call, with all its emotional urgency, than to years' worth of data that could be more predictive of how the case will play out in the long term if only they could make sense of it all.

At first, the algorithm looks more like a platinum stick than a nobleman fiddling with his measures. It seems to render judgments that are more consistent and less biased than human decisions. Look inside the black box, though, and the algorithm's decisions are shot through and through with human judgments. It derives its scores mostly from subjective decisions that have been repackaged in the Data Warehouse as snapshots of reality. The Data Warehouse receives information on people's use of social services only if they use publicly funded services. Private health insurers, addiction clinics, and therapists don't submit reports to the Data Warehouse, so people who have the means to get help without relying on public funding are much less likely to show up as suspicious creatures. Erin Dalton readily acknowledges this point. "We definitely oversample the poor. All of the data systems we have are biased. We still think this data can be helpful in protecting kids." (That said, the algorithm is no more biased against the poor than most child welfare agencies have always been, founded as they were by upper- and middle-class professionals to teach immigrants and industrial workers white middle-class norms.)

Because the algorithm relies so heavily on human decisions about families, it absorbs whatever biases, stereotypes, and cultural norms humans use to make their decisions. Dalton acknowledges this point, too. "All of the data on which the algorithm is based is biased. Black children are over-surveilled in our systems, and white children are under-surveilled. Who we investigate is not a function of who abuses. It's a function of who gets reported."

And who gets reported? Whoever is considered reportable by teachers, therapists, emergency room doctors, and other mandatory reporters, who use their own squishy criteria and intuitions to decide whether to make a referral to the department. Relatives and neighbors can use the hotline to settle a score or make trouble for someone they dislike. They can use the hotline with no accountability, because the department accepts anonymous calls. Anyone can report someone whose parenting doesn't conform to their own cultural and economic standards. The algorithm counts previous foster care placements as strong predictors of child maltreatment. Placement decisions are made by human service professionals and family court judges. As with the COMPAS algorithm in criminal justice, the Allegheny algorithm transforms official decisions made by government employees into risk traits of people the system has already swept up.

The screening tool, according to a recent evaluation, has improved the department's accuracy in identifying children at risk. But the algorithm is something of a self-fulfilling prophecy. It doesn't predict actual child maltreatment. It predicts two things thought to be warning signs. Prediction #1: How likely is it that the department will get another phone call about a family during the coming two months? Prediction #2: How likely is it that the child will be placed in foster care in the coming two years? When the tool gives a high score, the department will probably investigate the child's family. A case manager will talk with the teachers, doctors, therapists, police, neighbors, and relatives who interact with the family. Now that these people have been alerted to possible trouble, they'll be more likely to call the department when they have the slightest doubt about the family. Prediction #1 comes true. When more phone calls come in to the department and more concerns are raised, the case manager will be a little more cautious and a little more likely to recommend removing the child. Prediction #2 comes true. The more that screeners rely on the algorithm, the more accurate the algorithm will seem. No wonder, then, that the department has seen the tool's accuracy rate rise.

As Department of Human Services leaders considered adopting the algorithm, they knew it wasn't perfect. They knew it couldn't free itself from human bias and would sometimes make mistakes. They believed it was more accurate than humans, but they worried about the ethics of using a tool that might sometimes harm people with its mistakes. Before they implemented the screening tool, they asked two ethicists to conduct an ethical review of it.

For the ethicists, accuracy should be the paramount goal of screening and they believe the algorithm makes more accurate predictions than humans. "While it is true that predictive risk modeling tools will make errors . . . it is also true that they are . . . more accurate than any alternative, . . . even very good child protection professionals relying on their professional judgment and expertise." Given thier faith in mathematical tools, their conclusion comes as no surprise. "In our assessment," implementing the screening tool is "ethically appropriate. Indeed, we believe that there are significant ethical issues in not using the most accurate prediction measure." The Allegheny leaders got their ethical dispensation and the screening tool got its moral clout.

On close inspection, though, the ethical analysis isn't so clear-cut. Despite the authors' faith in mathematical tools, they devote most of their report to minimizing harms caused by the algorithm's inevitable mistakes. One suggestion: ensure that the professionals understand that the algorithm can "mis-categorize," that is, make mistakes. Huh? The algorithm predicts more accurately than professionals, but professionals should be trained to question its judgments. Not only is the advice contradictory. It flies in the face of some managers' view that humans should be trained to question themselves, not the algorithm.

Perhaps the biggest harm of mistakenly high scores is that they stigmatize people who've done nothing wrong. To reduce stigma, the ethicists again suggest training: "Emphasize that . . . risk scores are *only* risk scores and predictions" and that "individuals identified as at high risk must not be treated as though they have already been victims or perpetrators." In plain English, tell professionals not to stigmatize. If getting rid of stigma were that easy, it wouldn't be a problem, and besides, changing how professionals think doesn't change how people in the community think.

Some ethical issues arise from racial disparities in the data the algorithm uses. To deal with this problem, the ethicists compare protecting children and fighting crime. When police use racially biased criteria to stop and arrest people, their "interventions" are punitive and harm people who were wrongly classified. Child protection is different, the ethicists argue, because the interventions triggered by high scores are designed to identify people who need help and give it to them. "The fact that any intervention is designed to assist," they conclude, suggests that racial disparities in the child protection data aren't so troubling. But good intentions don't always equal good results. If they did, there wouldn't be much need for ethical analysis.

I went to college at the height of the Cold War and majored in Russian Studies. At some point, I learned about Soviet economic planning. The plans set production quotas for factories. If managers and workers didn't achieve the quotas, they lost their jobs. Coal mine managers were judged by the weight of their coal output. When they were short of coal, they filled freight cars with water before rolling them onto the scales. Textile factory quotas were specified in lengths of fabric. Weavers adjusted the looms to produce long, narrow strips. The moral of these stories was clear: central planning can't work and socialism is doomed to fail.

The stories were probably apocryphal but they impressed me with a different lesson: beware of perverse incentives.

In the early 1990s, right around the time Communism was declared dead, American business and policy schools began promoting a version of economic planning. They called it New Public Management, a name so empty and meaningless that no one could accuse it of being socialism or even government regulation. Eventually, the idea came to be called Performance Management and "pay-for-performance," which is the name I favor because it actually means something. It works exactly like Soviet economic planning: leaders set measurable goals, then reward people for meeting them or punish people for not meeting them.

Pay-for-performance jibes with the work ethic. No one should get paid just for showing up and going through the motions. Pay-for-performance jibes with common sense. If we want people to produce good results, we should pay them more for good results and less for poor ones. Pay-for-performance jibes with contemporary American political culture, too. It rings of individual responsibility, personal achievement, and entrepreneurial drive. For all its political appeal, though, pay-for-performance raises thorny practical and ethical questions.

On the practical side, the biggest hurdle is how to define our goals and measure whether we're meeting them. The Soviet stories come from a time of early industrialization, when leaders cared about producing physical things like coal and wool coats. Now most private and public sector production consists of intangible services, things like education, health care, economic well-being, security, and professional advice. It's much harder to define output for intangibles than for physical things, and even harder to measure quality. Still, pay-for-performance rests on an abiding faith that anything worth doing deserves to be measured. How else can we know whether we're succeeding?

Suppose we want to improve schools by rewarding good teachers. If—and this is a big if—the purpose of schools is to impart knowledge to students, then we can measure Mrs. Higginbotham's teaching ability by testing how much her students know at the end of a year in her classroom. With standardized tests, we can compare the quality of teachers and schools. Good teachers and good schools will produce students with higher test scores. And if that equation works, then tying teachers' pay to student test results ought to motivate them to teach better. That's the thinking behind George W. Bush's No Child Left Behind and Barack Obama's Race to the Top.

When I think back on the teachers who were most important to me, what they taught me wasn't anything that standardized tests can measure. My 7th-grade social studies teacher had us write reports on life in Colonial America. He suggested we include a preface but didn't give us any guidance except to say that a preface is about something more personal than research findings. When it came time to hand back our reports, he said that one student's preface exemplified what a good preface should do. Without divulging the student's name, he proceeded to read a portion of my preface aloud to the class. In that moment, Mr. Kingsley taught me that telling your readers why you care about the subject is a good thing, and that your emotional relationship to the material is as legitimate a topic as your ideas about it. That may be the lesson that made me the kind of scholar I am today, and it is certainly the one that makes me love what I do. I can't fathom how one would divvy up that lesson into countable units and test for it. And by the way, I don't remember a single fact about life in Colonial America from Mr. Kingsley's class.

Unfortunately, experiences like mine didn't inform current American education policy. Pay-for-performance has ruled the roost for over two decades. Students are subjected to mandatory standardized testing, primarily on math and reading. According to one survey, the typical student takes 112 standardized tests between preschool and 12th grade. Teachers' evaluations are determined by students' test scores. Their bonuses and contract renewals depend on test scores. In many places, principals and entire schools are judged on the basis of student test scores, and bad scores can shut a school down.

Teachers began protesting before pay-for-performance got off the ground. They knew what should have been obvious: what happens in students' lives outside the classroom has at least as much impact on what they learn as the formal teaching they receive. Some kids don't have parents who can help them with lessons. Some don't have enough to eat, a secure place to sleep, or a quiet place for homework. Some have to work after school or babysit their siblings. Some cope with parents' addictions and some have a parent in prison. Some are learning English from scratch. Some have illnesses and disabilities. Some witness violence, some are victims of violence. Some suffer from trauma that never leaves them. Let's just say the opportunities to study and learn aren't equal. Predictably, when teachers are judged by student test scores, teachers in what are euphemistically called "high-needs schools" don't do as well as teachers in wealthier school districts filled with professional parents.

Education reformers weren't naïve. Some had sympathy for teachers whose students face multiple hurdles before they walk into the classroom. The most ardent proponents of pay-for-performance, though, thought bad teachers were using poverty as an excuse. Eventually the reformers developed a mathematical model they claimed would eliminate all the extraneous factors and measure how much of a student's test score could be credited purely to a teacher's skills. This "value-added measurement" stands as a shrine to numbers. It creates imaginary students stripped of all their real-life troubles and gifts. The model assigns these ghost students "expected" test scores based on how they scored in the past. If a real student performs a lot better than his or her ghost, the student has "grown," and the teacher racks up quality points. A teacher whose students don't do much better than their ghosts gets a low rating. A teacher who teaches gifted students has a hard time scoring well because her students already scored at the top before she taught them. They have no room to grow.

If you have trouble following this model, so do I, so let's boil it down to a real person. Sheri Lederman teaches 4th grade in New York State. In New York, as in most states, teachers are rated as "effective," "developing," or "ineffective." (Get it? Good, Mediocre, and Bad.) Lederman's students consistently performed better than the state average on both the math and English tests, yet one year the model deemed her "effective" and the next year, with almost the same student test scores, she was rated "ineffective." Go figure. Lederman couldn't figure and neither could her district superintendent, who thought she was a stellar teacher. Lederman sued the state education department and won. Before the judge declared the ratings "arbitrary" and "capricious," the state had already changed its evaluation method.

Lured by federal money for states that agreed to adopt pay-for-performance, education officials got slap-happy with value-added assessment. In Washington, DC, even the custodians were evaluated partly on the basis of students' math and English scores. In Florida, half of a biology teacher's assessment was based on her students' scores on the state's reading test. In New York, art teachers' scores were largely based on how well students in their school did on the math tests. "I'm evaluated on a metric beyond my control," one art teacher blogged. Most teachers don't teach math or English, the two subjects required by the national education laws, but they're evaluated by those test results anyway. Reformers were so committed to evaluating quality by the numbers that they couldn't see the absurdity of grading teachers on students and subjects they never taught.

B. F. Skinner motivated rats to perform tasks by giving them food when they did what he wanted or electric shocks when they did something he didn't want them to do. Skinner and others in his tradition sometimes described their method as teaching through "conditioning." The context of the experiments suggests something more like brute coercion than education. The animals were trapped in cages, so they couldn't use their intelligence or their legs to find food and avoid danger. Typically, they were starved before their training sessions so they'd be highly motivated to do whatever was necessary to obtain food and avoid pain.

Pay-for-performance may not be as coercive as rat labs but it is a method of controlling people nonetheless. He who sets the numerical targets, measures performance, and doles out the consequences, exercises power. In pay-for-performance, numbers are the weapons of the strong. To the extent that people need income and employment, paying them or firing them according to their numbers pressures them into achieving good numbers as their primary goal.

But wait, you say. Pay-for-performance is the way of the world in most jobs (jobs without stratospheric earnings, that is). Why shouldn't teachers be subject to the same expectations? They should, but only if the performance measures are valid and fair. When people are trapped in a system of rewards and punishments they think is unrealistic or unjust, they resort to weapons of the weak: gaming the numbers and outright cheating.

Not long into the pay-for-performance regime, teachers stopped teaching anything that wouldn't be on the tests.

Principals and superintendents shifted resources away from art, music, humanities, history, social studies, and phys ed. Teachers devoted class time to drilling students for tests. Understanding and creative thinking went by the wayside. Some teachers leaked the test questions to students in advance. Some teachers corrected students' answers to yield higher scores. Some "lost" tests with low scores. School principals gave bonuses to teachers who raised student scores, no questions asked. Schools raised their test scores by artfully managing which students took the tests. A low-performing student could be suspended the day before the test. The school could force a special-needs student to transfer to another school by writing an individualized education plan that included services the school didn't offer.

No Child Left Behind and Race to the Top mandated states to achieve proficiency for 100 percent of their students by 2014. The laws also allowed the states to set their own proficiency levels. Surprise! Superintendents inflated student performance by lowering the thresholds for passing. In Washington, DC, where high schools were judged on their graduation rates, teachers gave diplomas to students who didn't meet the graduation requirements. In one high school, half the 2017 diplomas went to students who had missed more than three months of school. In Atlanta, Chicago, Dallas, New York City, and Washington, DC, dramatic student progress evaporated under outside scrutiny. The coal cars, it seemed, had been filled with water and the looms narrowed.

There were other perverse consequences. Pay-for-performance and the value-added assessment model are supposed to help schools in low-income neighborhoods and schools whose students face daunting psychosocial challenges. But the evaluation method doesn't fully account for the environmental factors, so teachers in those schools have lower ratings. The teachers get demoralized. Experienced teachers leave. The schools have a harder time recruiting good teachers. An evaluation system meant to help schools improve can send them on a downward spiral.

The metrics in educational pay-for-performance fail to measure the fullness of what teachers do. Teachers and principals find themselves trapped by Skinnerian taskmasters who control their livelihoods. Their reputations and self-esteem also depend on measuring up. The system puts them in a moral dilemma. Do they teach how they think is right for the children and risk losing their jobs and harming their own families? Or do they cheat however they can in order to live up to their personal missions and moral obligations? Before planners and managers blame people who game the system, they ought to think hard about whether their system has good values.

Any evaluation method relies on ideas about what's being evaluated and what counts as valuable. The value-added method embodies the long-discredited "container" view of education: teacher opens student's head and pours in knowledge. The name "value-added" conjures up an image of teachers depositing coins into a student's knowledge account. Most educators believe that good teachers do something else. They instill curiosity, creativity, self-confidence, ambition, and optimism. Those most precious affairs of the heart can't be sliced up and measured. That's not to say we shouldn't measure teachers' abilities and effectiveness, but judging them on the basis of student test scores is a bad idea.

One rock-bottom principle of criminal justice is that people shouldn't be held responsible for events over which they have no control. The value-added method holds teachers responsible for a zillion influences on student learning that teachers can't control. Math formulas are supposed to eliminate these factors—"control for them," in scientific jargon—so they don't influence a teacher's score. But scientific "control for" and human "control over" are two different animals. In the scientific method, controlling for extraneous influences doesn't make them go away; it just pretends they don't exist. Teachers don't live in that fantasy world. They must cope with the effects of poverty, violence, illness, and all the rest. Thanks to pay-for-performance, teachers can lose their jobs and schools can lose their funding because of poor value-added scores. Anyone who designs or uses evaluation measures ought to make "no punishment without guilt" a paramount ethical principle.

In a book called *The Tyranny of Metrics*, Jerry Muller traces pay-for-performance reforms in medicine, the military, universities, schools, policing, business, finance, philanthropy, and foreign aid. In every area, performance measures distort people's goals and make them aim to achieve good numbers instead of good results. Like Mickey Mouse and his magic broom in Disney's "The Sorcerer's Apprentice," humans have lost control of the tools we trusted to help us. But we should get clear about who's controlling whom. Numbers aren't tyrants; they're tyrants' weapons. Test scores, value-added scores, and proficiency levels can't force anyone to knuckle under. They gain their coercive power when powerful people use them to punish and reward. In No Child Left Behind and Race to the Top, national officials aimed the metrics at state officials, teachers, and principals like loaded guns. "Do as we say or else we'll fire you and close your schools."

Measures broadcast political messages. What we choose to measure signals what we think is important and frames how we think about problems. Our current way of evaluating teachers and schools lets the real tyrants escape unnoticed. Using poverty, gun violence, and addiction as "control variables" treats them as givens that can't be changed. With those messy things out of the picture, all eyes focus on the individual teacher and her 25 students in one classroom during one year. School reform comes down to asking teachers to do the best they can until society fixes its big problems. When politicians devote national attention to improving individual teachers, they pull resources and political pressure away from doing something about collective problems. It's not that we can't do both kinds of reform. We can and should, but we won't if we let narrow measures guide our efforts.

When people wield numbers to win arguments or justify their decisions, they don't casually send up a number to see if it flies. They make sure their numbers are fully armed with persuasive power. They wrap their numbers in flattering words. They vouch for their numbers. They brag about their numbers. They campaign for their numbers. If they want their numbers to influence how others think and act, they need to do more than present raw numbers. They need to practice the rhetorical arts to invest their numbers with authority.

James Madison wrote an elaborate essay to arm a fraction with enough firepower to blast it into the Constitution. He fabricated an imaginary courtroom argument before he declared three-fifths as the correct particle of personhood for slaves: "This is in fact their true character." With that simple sentence, Madison scored a rhetorical twofer: he pulled "facts" and "truth" onto the South's side.

You can play this game, too. Let's say you want people to use your numbers or your counting method. In other words, you want to find your numbers a job. Here's how to write a letter of recommendation for them.

Step 1: Brazenly assert, "My numbers are right." Use synonyms for "correct" to describe your numbers. Choose one or more terms from the following list:

True
Objective
Factual
Accurate
Valid
Reliable
Unbiased
Nonpartisan
Neutral
Scientific
Evidence-based

Data-driven
Fact-based
Research-based

Step 2: Pump yourself up as a trustworthy and credible source. Trust me, I've written hundreds of successful recommendations for my students. If you don't believe me, ask a marketing guru. The most influential part of an advertisement isn't the content but the source—who's telling the audience your product is the greatest thing since sliced bread. ("Nine out of 10 doctors prefer Camel cigarettes.") To enhance your status as a trustworthy source, use one or more of the following words to describe yourself:

Expert
Objective
Unbiased
Independent
Experienced
Renowned
Prize-winning
Leading authority

Don't worry that some of these words also appear in Step 1. You can't overemphasize your authority. For extra authority points, declare your commitment to high principles. Of course you are committed to:

Scientific rigor
Integrity
Honesty

Reaching the same goals as your audience (Throw in at least 2 of the following goals: justice, equality, efficiency, prosperity, savings, public safety, sustainability)

Step 3: Mention that your research or analysis was:

Rigorous
Comprehensive
Thorough
Exhaustive
Scientific
Data-based
Sophisticated
Innovative
An advance over previous research
Based on superior methodology (Don't say "method" when you can add 3 syllables and sound more sophisticated)
Methodologically strong (7 syllables—you must have a PhD)

And don't forget to add that your study used the most recent and best available data.

Step 4: Deny that fallible human judgment finds its way into your counting method. Instead of using people as the subject of your verbs, use some of the following abstract nouns:

Statistical analysis showed . . .
Statistical tools were used to . . .
Data were analyzed . . .
Analytics (or better yet, "data analytics") can reveal . . .

Models show that . . .
Quantitative assessment methods achieve accuracy . . .
Evaluation takes into account . . .

For example:

"Smart statistics are the key to fighting crime."

Or: "Value-added analysis offers the closest thing available to an objective assessment of teachers."

Or: "Value-added also may serve as the foundation for an accountability system at the level of individual educators" (that's "teachers" to you and me).

The passive voice is another excellent device for hiding human influence. For example:

"The data were coded . . ." (Translation: I counted Barack Obama as a black man with a white mother instead of as a white man with a black father.)

Step 5: Sprinkle your testimonials about your numbers with cautions to reassure doubters and show your humility. Use one or more of the following tactics:

Yes, my measure is imperfect, but it's the best we have.

Without a measure, we're operating in the dark, so we should keep using the one we have until somebody comes up with a better one.

Heavens, no, I don't intend my measure to replace human judgment. My measure is only an *aid* to human decision makers. They and they alone are responsible for their decisions.

> No one suggests using my measure as the sole measure of something so complex as [fill in the blank].

Put forth these caveats even if you know full well that people will use your measuring tool exactly as you say it shouldn't be used. As Tom Lehrer sang about the Cold War space engineer Wernher von Braun, "Once the rockets are up, who cares where they come down?"

My send-up is only half in jest. The mathematician John Ewing calls these rhetorical strategies "mathematical intimidation." Non-mathematicians make easy prey for these ploys. Think about the way arithmetic is taught. Teachers don't ask students to journal or write response papers about arithmetic. Arithmetic problems have a right answer. The student's job is to find it. We're taught to defer to the math experts.

Social issues aren't math problems. The danger of converting a social issue into a math problem lies in teaching citizens that there's one right answer and only mathematicians can find it. The question "What fraction of a slave is human?" does have a right answer, but it's a moral answer, not a number. "How good a teacher is Mrs. Higginbotham?" has many answers, depending on what you mean by good. Good has many meanings; few of them can be reduced to a number. Sometimes the right answer to a question is to challenge it.

5

How Counting Changes Hearts and Minds

To see how counting changes us, look no further than the Fitbit, the little wrist device that counts your daily steps. A Fitbit is about as precise and objective a measuring instrument as you can imagine. You can't fake it out. You can't make it think you've taken more steps by walking faster or shortening your strides. You can't fiddle with your Fitbit . . . but your Fitbit can fiddle with you. Unlike rocks or telephone poles, people think and feel. We have pride and shame. We want to look good to ourselves and others and we want to behave well and do well. We're vulnerable to the "Fitbit Effect": the process of measuring people changes the way they behave.

I've never had a Fitbit but almost everyone I know does and they all say their Fitbit makes them walk more than they otherwise would. It raises their daily step count by counting their steps. Partly that's because people who choose to wear a Fitbit have already decided they want to exercise more. They're motivated steppers and they use the Fitbit's numbers to prod themselves. Partly, the Fitbit's makers understand the human psyche. They program the Fitbit to set goals and give pep talks.

My friend Judy explained how her Fitbit changes her walking habits. The default goal setting is 10,000 steps a day. When Judy got her first Fitbit, she was "appalled at how few steps I walked on the days I didn't work," so she started taking long walks on those days to get her numbers up. By now, she almost always comes close to 10,000 steps without consciously trying. Sometimes she walks several blocks past her destination just to get in some more steps. She routinely stretches out her errands. "For example, yesterday I had a library book due. I could have returned it to our neighborhood branch about three-tenths of a mile from my condo, but instead I took it to the main branch, seven-tenths of a mile away."

The Fitbit has become Judy's coach. It expects her to walk at least 250 steps during every daytime hour. If she slacks off or, heaven forbid, sits in a chair reading for 50 minutes, the Fitbit vibrates and flashes a message such as "Only 86 steps to go" or "You have 10 minutes to get 250 steps." If and when she hits 250, her Fitbit tells her "Good job!" When she hits 10,000 steps in a day, it shoots off rockets. It tallies how many miles she has covered. It calculates daily and weekly fitness scores. Judy reeled off more goals and measures until my head was spinning.

"Here—go make daddy's Fitbit think he's exercising."

When Judy first described all the things her Fitbit counts and reports to her, I told her it sounded like a tyrant. She reacted strongly. "Trust me. I know tyrants and wouldn't be living with one. I truly experience my Fitbit as a low-maintenance personal assistant that *invites* me to stretch my legs once an hour, and to take a few more steps daily than I might have otherwise. It is my Jeeves; it serves me, not me it."

Whether the Fitbit is a tyrant or a Jeeves, counting can change the way we experience our lives. Erling Kagge loves to walk. He was the first person to trek on foot to the North Pole, the South Pole, and the summit of Mt. Everest. His mother and father took him on long hikes almost as soon as he could walk. In his lovely meditation, *Walking: One Step at a Time*, he recalls pestering his parents with the question "How much farther is it?" He surmises that the question might have been one of his first sentences, but now he thinks counting distance is a mistake. Forty days into his solo hike toward the South Pole, he writes, "I began to mentally calculate how much time I had left to go. Ten days times ten hours per day of walking would be 100 hours. At first I felt glad, but then this feeling changed. The next day, I wrote in my journal: 'The South Pole went from being a distant good dream to a mathematical point. This or that many hours left.'"

Counting changes our relationship to the world because in order to count, we narrow our focus to one or a few features of whatever we're counting. It all comes back to the basics of counting. Before we can count, we have to know what belongs in the category we want to count. We categorize by ignoring differences among things. Red fish, blue fish, fast fish, slow fish—they're all fish. While we count them, we stop noticing their differences. Dr. Seuss saw their differences because he *stopped* counting after 2. Counting heightens our awareness of the things we've decided to count and makes us ignore things we (or someone else) decided don't matter.

Something similar happened to food in the 1980s. Michael Pollan calls it nutritionism. "Where once familiar names of recognizable comestibles—things like eggs or breakfast cereal or cookies—claimed pride of place on brightly colored packages crowding the [supermarket] aisles, now new terms like 'fiber' and 'cholesterol' and 'saturated fat' rose to large-type prominence." Describing food by tallying its nutritional content changes our concept of food from a taste experience to a collection of countable nutrients—whatever *those* are. Most of us have no clue what vitamins, antioxidants, or proteins are. At best, we know them as dietary advice and advertising terms. The nutrient names on food labels don't conjure up any images but they still manage to evoke positive or negative emotions. Nutrition labels might as well be telling us how many fairies and gremlins live inside the jar. When we read labels, prepare meals, or choose our dinner from a restaurant menu, vague feelings of pride and guilt can crowd out our other associations with food.

Nutritionism, writes Pollan, redefines the purpose of eating as promoting physical health. Then he adds, "Food is also about pleasure, about community, about family and spirituality, about our relationship to the natural world, and about expressing our identity. As long as humans have been taking meals together, eating has been as much about culture as it has been about biology." Pollan gives new meaning to the expression "bean counters"—a derogatory term for people who count the Trivial instead of the Important because it's easy to measure the Trivial and hard to measure the Important.

"Self-tracking" is now something of a movement, with its own websites, meet-ups, research communities, and, not to be forgotten, device makers. Measuring weight, calories, blood pressure, and running times isn't new, but the variety of self-tracking has exploded with digital and Big Data capabilities. Now people are tracking their sleep, their moods, their stress, their leisure time, and their happiness.

Some research suggests that when people measure their own activity, such as walking, reading, or coloring, they enjoy it less than they do when they "just do it," as Kagge found on his trek to the South Pole. This finding isn't surprising in light of what we know about incentives as motivational tools. External rewards such as money, gifts, and prizes tend to diminish people's internal motivations. However, if you have set your own goal and you're counting to help yourself reach your goal, counting probably enhances your well-being, especially if you have some control over your measuring device. Judy can change the default goals built into her Fitbit. She can choose which measures to watch and how often she wants to look at them.

On the other hand, someone who's "outsourcing self-management from his own feelings . . . to an app" might "start to dwell less and less on what he wants and more on what the app advises." He might lose touch with his sense of self. Case in point: an editor for the website Quantifiedself.com monitors himself to learn how to "unwind more efficiently" and "optimize relaxation" on his vacations. "If my morning reading shows that I'm stressed, then I know not to push myself, maybe skip the big hike and leave more time that day for reading." Really? You can't just ask yourself what you feel like doing? And maybe pushing yourself for that big hike would help you unwind and even give you a high.

Clearly, people have different relationships with their Fitbits and other self-tracking devices. But no matter what kind of relationship they have, they're in a relationship and the relationship changes them, just as surely as our relationships with our families and friends change us.

The Fitbit Effect plays out in public opinion polls and attitude surveys, too. Just as Fitbits send suggestive signals to people they measure, pollsters ask suggestive questions to measure what people think. Why do pollsters want to measure public opinion? Because democracy is a form of government in which leaders follow the will of the people instead of the other way around. An inspiring ideal, democracy is, but how can leaders know what the people want? Elections are supposed to tell them, but they can't organize an election every time they want to know how the public feels about an issue.

George Gallup, the founder of Gallup Polls, believed opinion polls were the solution. Polls can take "the pulse of democracy." They're a "high-tech equivalent of the New England town meeting." Today's sophisticated opinion-research institutions share Gallup's democratic purpose. The National Opinion Research Center at the University of Chicago, for example, says it "taps into the public consciousness to provide government agencies and other organizations with the data and insights needed to understand and serve citizens in a world of vast and accelerating change." Tapping into the public consciousness isn't quite so simple as taking people's pulse or blood pressure, though, and even those relatively simple vital signs bend to the Fitbit Effect. Doctors call it the white coat syndrome. Your blood pressure goes up just because the doctor is watching and you're scared it will be too high.

When polls and surveys attempt to read people's minds, they also change people's minds. A survey, like a New England town meeting, is something of a public ritual rich in civics lessons. Participants in town meetings learn that rules and order are necessary for groups to be able to reach decisions. They learn that everyone has a right to be heard. They also learn that some people are more vocal than others, and more influential. If the moderator is any good, the citizens learn to stay on topic, to ask and answer informational questions, and to address each other's concerns. You won't find any of those civics lessons printed on the annual town agenda, though. Instead you'll find questions much like those on opinion surveys. "Should the town vote to appropriate $10,000,000 for highway maintenance?" The civics lessons are buried in the unspoken norms and procedures that govern the meeting.

Opinion research is now a branch of social science. It has its formal rules and best practices, just as every level of government does. It has its hanky-panky, too. Politicians can design polls and distort results to claim the support they want. That's the stuff of "how to lie with statistics" but I'm not so interested in that part. I'm interested in the flip side of surveys—not the answers people give but the lessons survey writers bury in the questions.

A man or woman comes to your door. "Good morning. I'm from the University of High Prestige and the National Institute for Important Research."

You think to yourself, "Never heard of either of them but they sure sound impressive."

The visitor continues: "We're doing a survey to learn more about how Americans think about issues facing our country. Would you be willing to help by answering some questions? Your answers will be completely confidential."

You'd like to help even though you're not sure anything you have to say can help anybody run the country better, so you agree to be interviewed. The interviewer asks many questions, then reads you some statements and asks how strongly you agree or disagree, or perhaps you "neither agree nor disagree."

When you're ready to take the survey, turn the page to begin.

Where would you rate blacks in general on this scale?
(The interviewer explains that 1 means you think almost all the people in the group are hard-working, and 7 means you think most people in the group are lazy.)

1. Hard-working
2.
3.
4.
5.
6.
7. Lazy

Where would you rate blacks in general on this scale?

1. Peaceful
2.
3.
4.
5.
6.
7. Violent

Where would you rate blacks in general on this scale?

1. Intelligent
2.
3.
4.
5.
6.
7. Unintelligent

Do you think that most white candidates who run for political office are better suited in terms of their intelligence to serve as an elected official than are most black candidates, that most black candidates are better suited in terms of their intelligence to serve as an elected official than are most white candidates, or do you think white and black candidates are equally suited in terms of their intelligence to serve as an elected official?

You'll be well into the interview before these questions about race come up and they might catch you by surprise. But don't worry. The survey is an equal-opportunity bigot. It asks you to rate whites, Asian Americans, and Hispanics on the same scales. It also asks you to rate Muslims and Christians as "patriotic" or "unpatriotic" on a scale from 1 to 7. And get this: the question about whether black and white candidates are equally suited for political office was asked for the first time in 2008—the first year Barack Obama ran for president.

I didn't make up these questions. I drew them from the American National Election Studies (ANES), because those studies are among the premier surveys on Americans' political views. The surveys are designed by top social scientists, mostly from the University of Michigan and Stanford University, and they receive support from the National Science Foundation. They are conducted every two years before and after each presidential and mid-term election. And the ANES uses what most social scientists consider the best methods for identifying group stereotypes and tracking racism and prejudice.

These questions invite respondents to generalize about entire groups based on skin color—not even skin color, really, but a *word* that supposedly describes skin color and a whole lot more. By asking questions laden with racial stereotypes, researchers put out the message that races are real things rather than amorphous human ideas about grouping people. The questions teach that generalizing about a group defined by a color word, a nationality, or an ethnic label is an acceptable way to think. Moreover, the questions about work ethic, violence, and intelligence perpetuate derogatory images of blacks that white Americans and Europeans have used to justify slavery and discrimination for centuries. If the questions were parents, I'd tell them they're modeling bad behavior to their children.

The next three questions invite stereotyping in a less obvious way.

> **I will read some statements. After each one, please tell me whether you agree strongly, agree somewhat, neither agree nor disagree, disagree somewhat, or disagree strongly.**
>
> **Irish, Italians, Jews and many other minorities overcame prejudice and worked their way up. Blacks should do the same without any special favors.**
>
> **Over the past years, blacks have gotten less than they deserve.**
>
> **It's really a matter of some people not trying hard enough; if blacks would only try harder they could be just as well off as whites.**

Unlike the earlier questions, these statements don't invoke such blatant racist clichés, but they do encourage people to judge entire groups as if all individuals in a group are the same.

Public opinion research is a two-way street. Surveys not only find out *what* people think about public issues; they also teach people *how* to think about public issues. You pick up the phone and someone speaking the Queen's English from an organization with the word "national" or "American" in its title asks you to rate blacks on a scale of 1 to 7 where 1 is "lazy" and 7 is "hardworking." Or a woman sporting a business suit and perfect teeth comes to your door, introduces herself as a researcher from a big-name university, and asks you whether you think blacks deserve their lower station in life because they don't try hard enough. What goes through your head? You might think, "Well, if *she's* characterizing entire racial groups, it must be an intelligent thing to do. I'd better come up with an answer if I don't want to appear stupid." Or you might think, "These questions make me furious. I know it's wrong to stereotype, but I was brought up to be polite, so I'll just keep my thoughts to myself and give a number in the middle." Or perhaps you're not so acquiescent and you say, "Get out of my face."

These questions make racial stereotyping seem legitimate and they encourage people to do it. Respondents have to buy into racial stereotyping in order to answer because there are no options for saying, "I don't think that's a reasonable question." Or "I don't think you can generalize like that." Or "I think it's just plain wrong to stereotype like you're asking me to do." Or "That's a racist question and I refuse to answer." The survey forces people to accept the questions without giving them any options that would reject the premises. The survey all but announces, "Racial stereotyping has our blessing."

One might argue that these questions don't encourage stereotyping; rather, they merely find out whether people already hold racist beliefs, just as a pathologist examines tissue samples to see whether they contain cancer cells. But reading minds isn't the same as reading tissue samples. The words and ideas researchers use can set off associations in respondents' minds and steer their thinking toward those thoughts. Social scientists call this process "priming." In the days before electric water pumps, to get water out of a hand-operated pump, you had to prime it first. You poured water from a bucket into the pump's spout as you moved the pump handle up and down. With only a little bit of water to get it started, the pump delivered more water so long as you kept pumping its handle.

Surveys, polls, and interviews are like pump priming. There's the researcher (a farmer) who wants to draw ideas (water) out of a person's mind (a well). The survey puts ideas in people's heads the same way a farmer puts water into the pump. Of course the respondent's mind already contains a lot of ideas, just as a well already contains a lot of water. The well of American culture contains images of blacks as lazy, stupid, and violent. Survey questions can activate ideas that are already there and call them up to consciousness. This is how the simple process of counting opinions can influence which opinions get expressed and counted.

To be sure, some surveys about race tap into both racist and non-racist ideas. For example, both the ANES and the National Opinion Research Center often use variations on this question:

> **Some people think blacks tend to have worse jobs and lower incomes than whites. Some people think whites tend to have worse jobs and lower incomes than blacks. Some people think there's not much difference. Which view is most like yours?**

If you agree that blacks tend to be worse off than whites, the interviewer will say:

> **I'm going to read you some possible explanations and I want you to tell me how important you think each is.**
> (If you say you agree that whites are worse off than blacks, you'll get the same options with the races reversed.)
>
> **Because whites have more in-born ability to learn.**
> **Because discrimination holds blacks back.**
> **Because blacks just don't work as hard.**
> **Because blacks don't get a chance to get a good education.**
> **Because blacks just choose low-paying jobs.**
> **Because government policies have helped whites more.**
> **Because God made people different from one another.**

On the surface, the question looks balanced because it includes both racist and non-racist explanations for inequality. All seven suggested explanations ask respondents to identify who or what is responsible for one race being worse off than another. The format allows them to give equal importance to all the explanations or to say some are more important than others. Three suggestions put the onus on blacks—they're not smart, they're lazy, and they prefer low-wage work. Three suggestions avoid blaming anyone by putting the onus on abstract nouns—discrimination, a chance at a good education, and government policies. These options prime respondents to consider systemic reasons for one race being worse off than another, rather than blaming the victims. The last answer is masterfully evasive *and* racist. It makes God responsible for everything on earth, including who gets good jobs and high income. It absolves both races. And without mentioning race, it taps into another age-old rationale for slavery: God meant for whites to dominate blacks.

Beneath the surface, the question sends a clear message. On the side of racist explanations, it serves up virulent racial stereotypes and a supposed religious justification for blacks being worse off than whites. On the side of non-racist explanations, it mentions vague social forces for which no one can be held responsible. Either way, racists aren't to blame. The two causal stories are unbalanced in their emotional power, too. The racist explanations mention human qualities that people can admire or disdain. Respondents can easily conjure up images of dense, lazy, unambitious folks they've known. The systemic explanations mention jargon words with less emotional resonance—government, education, and discrimination, although that last word packs a little more punch. Unless you and your family and friends are on the receiving end of oppressive social forces, real-life examples don't come so readily to mind.

Opinion surveys can nudge people's minds with the power of suggestion. They can also intensify any kind of conflict by portraying life as a competition for scarce resources and by making people feel threatened. Let's try some more questions.

> **What do you think are the chances these days that you or anyone in your family won't get a job or a promotion while an equally or less qualified black employee receives one instead?** (Possible answers are "extremely likely," "very likely," "moderately likely," "slightly likely," "not at all likely.")
>
> **How much influence do blacks have in American politics?** (Possible answers are "too much influence," "just about the right amount," "too little influence.") You'll be asked the same question about whites, Hispanics, and Asian Americans.
>
> **In general, does the federal government treat whites better than blacks, treat them both the same, or treat blacks better than whites?**
>
> **Do you think immigrants in the US strengthen our country because of their hard work and talents or are a burden on our country because they take our jobs, housing and health care?**
>
> **How important do you think the following is for being truly American?** (Possible answers are "very important," "fairly important," "not very important," "not important at all.")
>
> > **To have been born in America**
> > **To have American ancestry**
> > **To be able to speak English**
> > **To follow America's customs and traditions**
>
> **Next, I'll read some statements. Please tell me whether you agree strongly, agree somewhat, neither agree nor disagree, disagree somewhat, or disagree strongly.**
>
> > **Immigrants are generally good for America's economy.**
> > **America's culture is generally harmed by immigrants.**
> > **Immigrants increase crime rates in the United States.**

These questions couldn't be better designed to foment conflict. They're leading questions and they lead respondents to worry about losing out. They portray democracy as a zero-sum game, not as a cooperative endeavor to improve everyone's lives. They put respondents in a mindset of "us versus them." Do politicians listen to one group more than another? Does the government favor one group over another? Does one group get more than its members deserve? Does one group lose out in the competition for jobs? The questions almost tell respondents which suspect in a lineup is the guilty one.

The questions about immigrants portray American life as a grand contest between immigrants and natives. To see why, try this thought experiment with questions that I made up:

Do you think people born in the US strengthen our country because of their hard work and talents or do they burden our country because we have to pay for their education and health care?

I'll read some statements. Please tell me whether you agree or disagree.

Are people born in the United States generally good for the economy?
Is American culture generally harmed by people born in the United States?
Do people born in the United States increase crime rates in the United States?

These questions probably seem bizarre because it would never occur to you to generalize in these ways about everyone born in the U.S. By asking such questions only about immigrants, the surveys portray natives as one homogeneous body politic and make immigrants seem like foreign agents with mysterious powers to harm or help the native body.

Here's another thought experiment: The survey asks what criteria you use to count someone as "truly American." It provides several ideas for you to consider. No matter what you answer, how will you think about people who don't have what you believe it takes to be a true American? Is someone who doesn't speak English well a fake American? A disloyal American? A traitor? The phrase "truly American" invites respondents to imagine an "other," and whatever that other is, it's not good, because everyone knows that "true" is right and "true" is good.

Survey research on prejudice, racism, and political conflict has good intentions. It aims to understand the psychological basis of political hostility and racial animus. For years, I taught some of this research in my courses as if the shocker were that so many people openly admitted to racist views. Then, in 2011, I taught a course in Denmark on American race politics. One evening as I was preparing for class, I wondered, as I always do, what my students would make of the readings I'd assigned. Suddenly, seeing through young Danish eyes, I felt shocked that academics would *ask* questions like these, never mind the answers. Since then, whenever I teach about these racism surveys, I ask my classes what they think of the survey questions. My students at first share my dismay, but eventually, someone comes to the survey's defense.

In one class, a delightful Nigerian woman made the case with an impish twinkle in her eye and an unforgettable question: "I don't mean any disrespect, Professor, but how else are we going to find out whether white people think Africans live in trees if we don't ask them?"

I see the point, and the researchers certainly do, too. But if we're going to ask people if they think Africans live in trees, let's also give them opportunities to express incredulity and outrage, or at least let them disagree with the premises of the questions. Let's examine the hidden curriculum of measurement instruments. Let's write questions with an ear to what they hint at as well as to what they say. In the process of measuring conflict, we can also design measures that help people come together.

For example: In addition to asking whites how likely it is that they'll be scumped out of a job by a black person, we could ask:

> **Do you know anyone who you think probably got their job more because of their connections than their qualifications?**
>
> **In the place where you work, how much do you think personal relationships with higher-ups influence who gets raises and promotions?**

Questions like these might put a chink in the belief that everyone earns whatever goodies they have solely by their own hard work. These questions hint that the world of work isn't a perfect meritocracy, without suggesting that one group is to blame. They reframe the conflict from one between demographic groups to one between two standards of distributive justice: merit and personal loyalty.

We could also ask questions that illuminate how everyone gets help from government.

> **Have you or anyone in your family ever received help from government?**

Most people will probably say no because they don't know the half of what government does for them. ("Keep government out of Medicare," some anti-Obamacare signs read.) It's time to probe:

> **Did you or anyone in your family go to public school?**
>
> **Do you have running water in your home? If so, who provides it? Is it safe to drink? Who makes sure it's safe?**
>
> **When you buy groceries, do you trust that the food is safe to eat?"** If yes, ask, **Who do you think inspects food to make sure it is safe and recalls contaminated products?**
>
> **Have you or anyone in your family ever received payments from any of the following government programs: Social Security, Medicare, Medicaid, or Disability?**

These questions highlight the benefits of cooperation instead of the threat of competition.

Surveys about immigration could send more positive messages than they typically do. For example:

> **I'm going to read a list of ways immigrants contribute to our communities. For each one, do you agree, disagree, or neither agree nor disagree?**
>
> > **Immigrants tend to work very hard because they are trying to make a better life for themselves and their children.**
> >
> > **Many immigrants do jobs that long-term residents don't want to do.**
> >
> > **Many immigrant parents encourage their children to study hard and do well in school.**
> >
> > **Many immigrants take jobs beneath their qualifications in order to give their children a chance to climb higher.**

These questions plant positive images of immigrants. Even if respondents disagree, they will have been exposed to an alternative idea.

Psychologists have been experimenting with techniques to reduce unconscious bias against blacks and other stigmatized groups. The techniques that work best give people short narratives to read before they take a test for "implicit bias." The narratives portray blacks in a positive light. Some narratives contain only a sentence that mentions names of famous, celebrated blacks and notorious, bad-guy whites, yet even they reduce bias. The most effective technique for reducing unconscious bias puts you, the reader, into a story as a victim. The storyline says you are being attacked by a white person and a black person comes to your aid. Apparently, portraying blacks as helpers and allies reduces negative feelings about them.

Building on these techniques for reducing racial bias, here are some questions surveys could ask about immigrants that might help reduce anti-immigrant sentiment.

Have you ever been in any of the following situations:

> **An immigrant aide, nurse, or doctor took care of you or someone in your family?**
>
> **An immigrant building janitor helped you carry a heavy package or fix something in your apartment?**
>
> **An immigrant Uber driver went beyond the job for you?**
>
> **An immigrant teacher or classroom aide worked especially well with your child?**

If you could have dinner with any of the following immigrants to the U.S., who would be your top choice?

Sergey Brin, cofounder of Google (from Russia)
"Big Papi" David Ortiz, Red Sox player (from Dominican Republic)
Isabel Allende, author (from Peru and Chile)
Trevor Noah, late-night comedian (from South Africa)
Natalie Portman, actress (from Israel)

The point of putting a question like this on a public opinion poll isn't to find out what people think. Politicians don't care who's popular with their constituents. The point is to link immigrants with positive associations and change thinking by the power of suggestion.

Public opinion surveys are the best tool we have for transmitting citizens' concerns and desires to leaders. Ideally, surveys pose thoughtful questions and enable citizens to give considered responses. In practice, surveys more often pose simplistic questions and encourage simplistic thinking—or no thinking at all.

Let's try a few more items from actual ANES surveys. I'll present a statement and ask for your opinion before I weigh in with my two cents.

The world would be a better place if people from other countries were more like Americans. Do you agree strongly, agree somewhat, or neither agree nor disagree?

If I were taking this survey, I'd want to ask the interviewer, "What exactly do you have in mind?" The world would be better if Danes bicycled less and carried more automatic weapons? If the Japanese consumed less seafood and more sugar and fat? If Ugandans dressed more like Americans (whatever that is)? And what do you mean by "better"?

> **Our country would be great if we honor the ways of our forefathers, do what the authorities tell us to do, and get rid of the 'rotten apples' who are ruining everything. Do you agree strongly, agree somewhat, neither agree nor disagree, disagree somewhat, or disagree strongly?**

This is a rant, not a statement that can be analyzed. How do you answer if you're someone who thinks before you shoot off your mouth? If I came upon this statement in a student paper, I'd be all over the student to define greatness. I'd ask, "Which ways of our forefathers—bravery or slavery? Who do you have in mind for rotten apples?"

> **'Fracking' is a way to drill for natural gas by pumping high-pressure fluid into the ground. Do you favor, oppose, or neither favor nor oppose fracking in the U.S.?**

This question provides respondents with a definition of something they may not have heard of. So far so good. But then, without giving them any further information, it asks them for their opinion. In one fell swoop, the question sends three messages: One: fracking is controversial. Two: you ought to have an opinion about it. Three: you should make a snap judgment based on nothing.

> **Do the health benefits of vaccinations generally outweigh the risks, do the risks outweigh the benefits, or is there no difference?**

There is a body of scientific evidence on vaccines. Not surprisingly, the evidence differs from vaccine to vaccine and for different kinds of people (healthy adults, infants, seniors, those whose immune systems are weak). Vaccination creates health benefits for communities as well as individuals, because people who get vaccinated against a disease are less likely to get sick and spread the disease to others. By posing the vaccine question in such a sweeping way, the survey tells people they don't need to make distinctions to evaluate claims.

Answers to these questions are supposed to convey to political leaders how the public thinks. But the questions also convey some lessons about democracy to the public whose opinion political leaders seek. The questions tell citizens that politicians are interested in people's opinions regardless of how well grounded they are. Most teachers try to convey the opposite lesson. We discourage simplistic, uninformed, knee-jerk answers. We ask students to define terms, question assumptions, and make distinctions. We ask them to support their opinions with reasons and evidence. We challenge them and downgrade them when they make sweeping generalizations of the kind that opinion surveys often do.

By revealing some of the hidden lessons in surveys, I'm not impugning all opinion research. I deliberately selected the most flagrantly racist questions I could find, and the ones that make the most outlandish generalizations. I chose them not because they're typical but because they so clearly demonstrate the Fitbit Effect in public opinion surveys. They call to mind a lesson I learned early in my teaching career: if you don't want bullshit answers, don't ask bullshit questions.

To see some really cool Fitbit Effects, it's worth one more trip to the Census Bureau. The Census Bureau counts people and, in the process, it can send them scattering or it can reel them into its net. Donald Trump attempted to capitalize on the Fitbit Effect by adding a citizenship question to the 2020 census. "Is this person a citizen of the United States?" Trump had campaigned on a promise to curb immigration, and once in office he threatened to deport as many immigrants as he could. He knew what asking a citizenship question would do to immigrants.

Trump, it turned out, was taking advice from some Republican strategists who were scheming how to strengthen the Republican Party's grip on Congress. They saw a citizenship question as the best way to weaken the Democratic Party. Why? Because immigrants tend to vote Democratic, and because in the Trump era, asking a citizenship question on the census would be like firing tear gas into immigrant communities. People would run from the census. The question would result in undercounts of population in the cities and counties where immigrants tend to live. With population down in those areas, electoral maps would have to be redrawn. Republicans would gain more seats in Congress.

If anyone needed proof that a citizenship question would have this effect, the Census Bureau already had it. The bureau conducts focus groups and pretests during the years leading up to each decennial census. Already in 2017, before the idea of a citizenship question became public, these pretests uncovered unprecedented resistance to talking with Census Bureau interviewers. Hispanics, Asian Americans, and Muslims feared the Census Bureau would send their information directly to ICE, the Immigration and Customs Enforcement agency. The next thing they knew, they'd be deported or find out they'd delivered someone else into the cold clutches of ICE. "The possibility that Census could give my information to internal security and immigration could come and arrest me for not having documents terrifies me," one Spanish speaker said. The scuttlebutt, field workers learned, was "Don't open the door to census takers."

All this fear and evasion was happening *before* plans to add a citizenship question hit the news. Once Trump's plans became public, Census Bureau experts weighed in against it. They said the question would undermine the accuracy of the head count. If the Census Bureau values anything, it's accuracy. It also values its reputation as scrupulously professional and nonpartisan. The experts worried that a citizenship question would damage the bureau's reputation, as well as the count.

The Trump administration withdrew its demand for a citizenship question just as the Supreme Court was about to rule on the case, but the damage was already done. Anti-immigrant rhetoric and threats had put the fear of ICE in people. In that political climate, census officials would have a tough job convincing immigrants and their relatives to answer the 2020 census. The bureau tried to give them some assurance in its publicity flyers: "Your responses to the 2020 Census are safe, secure and protected by federal law. Your answers can only be used to produce statistics—**they cannot be used against you in any way.**" Time will tell whether immigrant communities bought the promise, but I suspect that the Census Bureau will be accused of undercounting and that the 2020 census results will be challenged.

When the Hispanic question was added to the census in 1980, it set in motion a dynamic quite the reverse of "run-for-the-hills." Beginning in the 1960s, people with ties to Spanish-speaking countries not only wanted to be counted but demanded to be counted in a way that would help them get social services and better political representation. They also wanted to be counted in a way that reflected their own sense of who they were. The Census Bureau's decision to count Hispanics played an important part in creating a Hispanic identity.

Before the 1960s, there was no Hispanic or Latino identity. The three main Spanish-speaking groups in the U.S. came from Mexico, Puerto Rico, and Cuba. They lived in different parts of the country and didn't interact. Wherever they lived, they were usually segregated, discriminated against, and in the case of Mexicans, denied the vote. As more people came from Central and South American countries, they migrated to cities and smaller towns throughout the U.S., but still tended to cluster by national origin, as immigrants do everywhere. They identified primarily with their country of origin, not so much as Spanish speakers or Spanish origin. When the Census Bureau lumped all these people from different countries into one large group, the new category answered many political needs.

In the 1960s, the civil rights movement and race riots had pushed racial and ethnic inequalities to the top of the domestic agenda. Government responded with laws and programs to benefit disadvantaged minority groups: blacks, Hispanics, Alaska natives, and Asian-Pacific Islanders. The Civil Rights Act (1964), the Voting Rights Act (1965), the Equal Employment Opportunity Act (1972), and other federal aid programs needed data on these groups in order to carry out their corrective missions. With political and social benefits to be had, Hispanics had high stakes in being counted. The bigger their numbers, the bigger the potential benefits. Leaders of several national-origin groups, especially Mexican Americans, saw the political advantage of joining forces to boost their numbers. White political leaders saw political advantage in the new category, too—a new voting bloc they could woo.

Meanwhile, the Census Bureau was under pressure to do a better job of counting blacks and Hispanics, because it had undercounted them in the 1970 census. Hispanic leaders had sued the Census Bureau to prevent use of the 1970 results. The suit was unsuccessful, but the negative publicity called into question the census's legitimacy. The Census Bureau wanted a full count of Hispanics in the 1980 census in order to restore its credibility.

This convergence of political interests led to a new question in the 1980 census: "Is this person Hispanic?" The question has remained on the decennial census ever since, although in 2000 it was reworded to include three labels: "Is this person Spanish/Hispanic/Latino?" Remember our preschool counting lessons from Chapter 1. A triangle is only a name until someone teaches you the rules for what counts as a triangle—or in this case, for who counts as Hispanic. The Census Bureau would have to become that teacher.

The Hispanic ethnicity question relies on people to identify themselves as Hispanic. That means checking off "yes" or "no" after the question "Is this person Hispanic?" If a person checks yes, the question asks them to write in a more specific description, such as Mexican American or Dominican. Vincent Barabba, the director of the Census Bureau appointed by President Nixon in 1973, worried that people wouldn't know what "Hispanic" meant. He also worried that Hispanics might not send back their census forms. He knew that immigrants, especially undocumented ones, feared punishment, deportation, and discrimination. (Their fear wasn't baseless. The 1930 census had included Mexican Americans as a separate race category. A few months after the census, President Hoover deported thousands of Mexican Americans, including many who were citizens.)

How, then, could the Census Bureau convince people to answer the census and to identify themselves as Hispanic? Barabba sought help from the people he hoped to teach. He created two advisory committees, one for blacks and one for Hispanics. Instead of staffing them with the usual statisticians and demographers, he appointed black and Hispanic community leaders. He invited Hispanic civic group leaders and Spanish-language media executives to a meeting, where he asked them to promote the census and encourage their followers to answer yes to the Hispanic question. The civic leaders were all in, because their communities could gain electoral power, legal protections, and federal resources from bigger numbers. The media executives were all in because they could gain more advertising revenue by showing they had bigger audiences. The one Spanish-language television network at the time created commercials and weekly features about the new Hispanic question, and when the 1980 census forms were distributed, the network set up a hotline to help people fill them out.

The Census Bureau's strategy to boost the Hispanic count worked much like a Fitbit. The census itself was the Fitbit's electronic step counter. The promotional campaigns were the Fitbit's goals and pep talks. The Hispanic or Latino identity caught on, not only in the census but also in everyday conversation. By 2002, over 80 percent of people whose national origin was in Latin America, Puerto Rico, or the Dominican Republic said they sometimes referred to themselves as Hispanic or Latino. A decade earlier, fewer than 20 percent of people whose national origin was in Mexico, Puerto Rico, or Cuba had described themselves as Hispanic or Latino. The census wasn't the only cause of this change, of course. Since the 1970s, almost every government agency and private organization has collected race and ethnicity data on people they serve. Schools, hospitals, insurance companies, businesses, nonprofits—they all want your identity. These ubiquitous questions shape the public's understanding of racial and ethnic categories at least as much as the census. But the Census Bureau took on a teaching mission with more zeal than any other organization and its mission continues.

Census officials have never been satisfied with how Hispanics answer the race and ethnicity questions. The Hispanic question is separate from the standard race question, "What is this person's race?" The race question doesn't ask for a simple yes or no as the Hispanic question does. It asks people to classify themselves as white, black, Asian, Native American, or any of about 15 other options. People can check as many boxes as they wish and if they don't see themselves as fitting into any of the options, they can check "Some other race" and write in a name for their race. New in the 2020 census, people who say they're "white" are asked where their ancestors are from.

Woe to the teacher who doesn't give students clear counting rules. To the Census Bureau, Hispanic is an ethnicity, not a race and not a nationality. But among people trying to make sense of the Hispanic and race questions, utter confusion reigns. In the 1980 and 1990 censuses, over 40 percent of people who said they were Hispanic checked "Some other race" in the race question. When offered the chance to specify their race, many wrote "Hispanic" or their family's country of origin. And if the numbers from the 2010 census are to be believed, "Some other race" is now the third-largest "race" in the U.S., after white and black.

The Census Bureau might be a Fitbit on steroids. It simultaneously counts and shapes the count. Ever since Barabba enlisted the help of Hispanic leaders for the 1980 census, the bureau has actively tried to influence how people answer the Hispanic and race questions and to discourage them from answering "Some other race." The bureau tests different ways of wording questions to find out which wording elicits the fewest "Some other race" answers. For the 2010 census, the bureau inserted an explicit instruction into the race question: "Hispanic is not a race." Despite the instruction, almost 37 percent of Hispanics declared themselves "Some other race."

Sociologist Clara Rodriguez has studied how Latinos (her preferred term) interpret race and ethnicity labels and why they answer census questions as they do. Many Latinos refuse to place themselves in an American race category because they think of race as something to do with their culture and traditions—not their biology or their blood. Some refuse to call themselves by American race labels because they understand the consequences of being white or black in the U.S. A Puerto Rican woman told Rodriguez, "The only time I respond that I am 'white' on a questionnaire is when I'm applying for a mortgage or a loan." A Dominican man told Rodriguez why he had chosen "Some other race": "By inheritance I am Hispanic, but to white America, if you're my color, you're a nigger." Being counted in the census isn't a passive process. The census is more like a fencing match between two sparring partners who influence each other's moves.

National censuses the world over divide populations into categories. Different countries at different times count race, ethnicity, religion, language, caste, and indigenous status. The counting categories may vary but they have one thing in common. They don't grow from the earth or rain down from the skies. They grow from human beliefs about who belongs in which category and how people in different categories should be treated by people in other categories. Often, laws establish rules and rights for different categories. Censuses convert the mix of beliefs, practices, rules, and rights into artificial categories with seemingly clear boundaries.

Certainly social divisions exist before censuses measure them, and certainly social practices and laws lead a census to count the divisions that matter politically. But according to a study of 31 countries, those that counted politically important ethnic divisions in their 2000 census were more likely to have outbreaks of ethnic conflict and violence later in the decade. The researchers don't argue that censuses *cause* conflict. Rather, they suggest, censuses freeze ethnic divisions into cold numbers and publicize who's up and who's down. Quite plausibly, ethnic counting fans the flames of preexisting conflict.

Whatever counting people does, it doesn't leave them unchanged. It can change how they see themselves, how they see others, and everything that follows from who we think "we" are and who we think "they" are.

Often we teach kids to count by making a game of it. As the last step in the game, the child announces a number—let's say, "9." Whatever number the child gets, it is more than a number. It is a powerful declaration: "There are 9 of these things here in this place." So what? you ask. What's so powerful about announcing your tally? When people believe something is rare or nonexistent, asserting that you've seen many "somethings" can persuade them to change their minds.

In 1992, Barry Scheck and Peter Neufeld, both lawyers and former public defenders, founded the Innocence Project at Cardozo Law School in New York. Their mission was to find innocent prisoners and get them exonerated. They focused on using

DNA evidence to prove innocence, but they also tried to figure out what was happening in the criminal justice system to allow innocent people to be falsely convicted. Innocent people, they found, were being convicted on the basis of false eyewitness testimony, coerced confessions, misidentification in lineups, misbehavior by police and prosecutors, and incompetence on the part of defense attorneys. Scheck and Neufeld's idea caught on. By now, there are about 50 independent innocence projects and advocacy centers around the U.S., all dedicated to the same mission. Innocence work has become a veritable social movement.

During the early years of the innocence movement, most lawyers, prosecutors, and judges in the criminal justice system insisted that false convictions were extremely rare. Some even claimed that false convictions *never* happened. Innocence advocates kept count of exoneration cases—cases in which a person who'd been convicted, sentenced, and imprisoned was later proven innocent. At first, the exoneration numbers mounted slowly, one in this state, three in that state. It took more than a decade for the projects to identify 100 wrongfully condemned people. Despite the small numbers, though, the movement gained some traction. In 2001, Illinois governor George Ryan issued a moratorium on executions in his state.

How could such small numbers trigger such a big change? For starters, innocence advocates used mounting numbers to chip away at claims that false convictions are exceedingly rare. The exoneration numbers did seem to change some minds. In the mid-1980s, criminal justice professionals estimated that false convictions happened in fewer than half of one percent of serious felony cases. By 2008, the professionals believed false convictions occurred in as many as 3 percent of serious felony cases. We can't know why their estimates rose, but it's a good bet that hundreds of widely publicized exoneration cases had something to do with it.

The innocence movement boosts the power of small numbers by blending them with stories. Here's how that blender works. In 1998, some innocence activists organized a conference in Chicago to highlight flaws in the capital-punishment system. One of the organizers had the idea to invite the 74 people who had thus far been exonerated. "We don't need to get 'em all," he said. "We just need to get some respectable number." The organizers got more than 30 to attend, but the numbers didn't matter so much as what the exonerees said and what they represented. They went to the stage, one by one, to give a somewhat scripted speech. As a journalist who was at the event recalled, the speeches went something like this: "My name is Jim Burrows. The state of Illinois sought to kill me for a murder I did not commit. I was put on death row in 1989. I was released in 1994. If the state had its way, I'd be dead today." Each exoneree placed a sunflower in a vase "to symbolize the life they had regained." They sat together on the stage. One exoneree called their collective moment on stage "a living graveyard." The conference grabbed international media attention, mobilized death penalty abolitionists, and galvanized the innocence movement.

The movement is powered by the same electric blend of numbers and stories that ignited the Chicago conference. Each new exoneration case came with a human story, and innocence advocates made sure to tell it. Exonerees and their families talked about the impact of false conviction on their lives. Newspapers, TV, and podcasts publicized the stories. Conventional wisdom holds that stories and photos of individual victims are more effective than statistics for creating empathy and moving people to act against suffering. By linking each new "1" in the exoneration tally with a human drama, innocence advocates bridged the psychological gap between statistics and emotion. And by showing each exoneree together with family members, the advocates effectively multiplied every "1" into a bigger number.

Advocates amplified small numbers in other ways. Each exoneration, they said, added up to 2 people—an innocent person sitting in prison and a guilty criminal on the loose. In many exoneration cases, DNA evidence also identified the real perpetrators. Many of them had committed more crimes while they were free. Innocence advocates turned these numbers into a public safety problem. Every false conviction, they pointed out, increased the chances that more crimes would be committed by the real perpetrator.

Each exonerated convict was only a single case. By any scientific standard, one case is merely an anecdote and could easily be a fluke. But when placed against the American standard of justice—no punishment without guilt—each wrongfully condemned person became a resounding indictment of the criminal justice system. That story amplified the numbers, too.

The National Registry of Exonerations keeps a running tally of exonerations since 1989. As of March 2020, there were 2,570 exonerations. The registry's website features an interactive graphic where you can not only see national numbers but also click on any state to find out how many people in it have been exonerated. The display also tallies how many years exonerated people have languished in prison as a result of false convictions. By March 2020, that number was 23,064 years—almost 10 times the number of exonerations. The "years lost" number transforms exoneration cases into living, breathing, feeling people whose life stories unfold in prison. Counting years lost in prison is yet another way advocates amplify the numbers to heighten their impact.

National Registry of Exonerations

2,570 Exonerations Since 1989	**23,064** Years Lost—Total	**9.0** Years Lost—Average Per Case

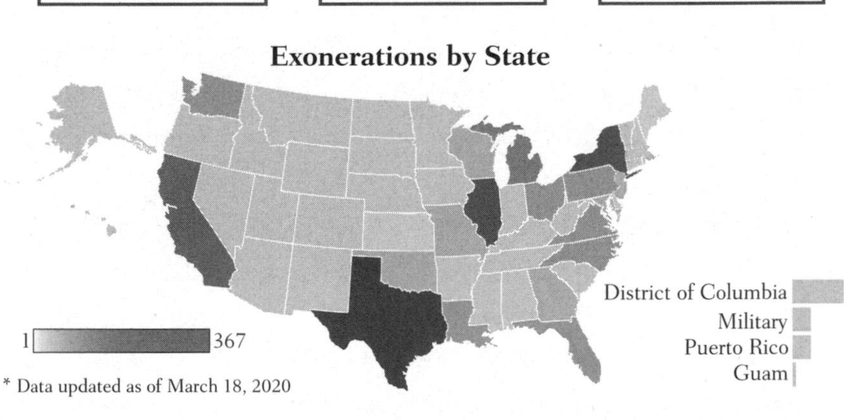

The National Registry acts as a kind of Fitbit for innocence advocates. People don't wear a Fitbit unless they're already motivated to exercise. The Fitbit prods them to do more by showing them numbers, setting goals, and giving them a pat on the back when they reach a goal. Innocence advocates go searching for false convictions because they're already motivated to do justice work. The Registry tracks their progress in exoneration numbers and shows them how much more work there is to do. And the Registry pats them on the back with "milestone reports" to publicize their political successes. In 2019: "Indiana Becomes 34th State to Pay Exonerees Compensation for Wrongful Incarceration." (Money can't make up for lost time and opportunities, but it can help exonerees to build careers, establish homes, and save for retirement.) Another milestone reached in 2019: "Oklahoma Becomes 25th State to Require Recording of Interrogations." (Recordings enable defense lawyers to determine whether the police deceived suspects or coerced them to confess.) These milestone numbers encourage advocates and build momentum for other states to get on the right side of history.

I asked Samuel Gross, founder and longtime director of the National Registry, what he thinks of my Fitbit analogy. He was quick to say that he started the Registry as a scholarly enterprise and its goal remains research and policy analysis. The Registry's researchers don't stop at totals. They look for patterns in the numbers, such as racial disparities in false convictions and the most frequent causes of false convictions. Nevertheless, Gross agreed that the numbers probably motivate advocates and help them win support. He also emphasized, as I do, that without stories, the numbers would have less clout. "We now often say we do two things at the Registry—tell stories and count things—and they are equally important."

Something magical happens when we count. We count to learn what's happening in our world and to gain control over our lives. Pretty soon, the process of counting changes the world and changes us. We can't find out what people think unless we ask them, yet the way we pose our questions affects what they think, or at least what they tell us they think. We can't find out how many people live in our country or anything about them unless they come forward to be counted and tell us about themselves. Yet who comes forward and what they say about themselves depends on who's doing the counting and why the "counters" want the numbers.

Counting can make ghosts become real. In 1923, a famous judge with the ghostly name of Learned Hand declared that there was no such person as "the innocent man convicted." That idea, he said, was merely a ghost who haunted the courts. People should stop worrying about it. A few decades later, a tiny band of people passionately devoted to justice thought they saw ghosts and started counting them. Their visions inspired more people to search for ghosts. The number of ghost sightings grew and grew. More and more people started believing in ghosts. Reformers changed laws and procedures to restore ghosts to their humanity and curtail ghost production.

We make numbers and our numbers make us.

6

The Ethics of Counting

Counting is a lot like being a judge. Before we can *count up* "somethings," we have to decide what *counts as* a something. Before judges can decide which laws and precedents to apply, they need to know how the case they're now considering is like other cases that courts have already decided. Is it *this* kind of a case or *that* kind of a case?

One of the most important American civil rights cases involved a bakery that specializes in wedding cakes. The bakery owner had refused to bake a wedding cake for two gay men. The men thought they'd been discriminated against and went to the Colorado Civil Rights Commission. The commission ruled that the baker couldn't discriminate against gays. The baker appealed his case all the way to the Supreme Court. He said baking wedding cakes is an artistic activity. He had nothing against gays, he said, but his religion forbids gay marriage. If he were required to bake a cake for a gay couple, he'd be forced to express his approval and to go against his religious beliefs. "I'm being forced to use my creativity, my talents and my art for an event—a significant religious event—that violates my religious faith," the baker told a reporter.

The dispute turned on how to *count as*. Is a wedding cake shop like a restaurant because the owners serve food? Or is it more like an art studio because bakers express themselves through their culinary designs? If a wedding cake business counts as a restaurant, then civil rights laws require the owners to serve all comers without discriminating. But if it counts as an art studio, then the First Amendment prohibits government from forcing people to express beliefs.

When judges decide such hard questions, they're not counting in a numerical sense, but they are making comparisons in exactly the same way we do when we count. In this case and others like it, both ways of seeing the issue can be right. From one perspective, the case is about people's right not to be discriminated against. From the other perspective, the case is about people's right to freedom of speech and religion. American law recognizes both rights. When the two rights collide, judges must *count up* to decide which one is more important and should outweigh the other.

Judges' decisions about *counting as* and *counting up* have momentous consequences beyond a single case. If bakers can say that their services count as expressions of religious beliefs, why not photographers, architects, lawyers, EMTs, doctors, teachers, or even judges? From there, it's a short step to saying that professionals and government officials can't be required to provide their services to blacks, women, gays, Jews, or Muslims, so long as they claim that serving these groups violates their religion. At that point, antidiscrimination law loses all its teeth and becomes unenforceable. That's how counting decisions can knock down a pillar of justice.

To end your suspense, the Supreme Court kicked the can down the road. It ruled in favor of the baker without deciding how to classify a wedding cake business. Instead, the Court rebuked the Colorado Civil Rights Commission for showing hostility to religion when it decided in favor of the gay couple. The Supreme Court's decision applies only to this one baker. But the Court won't be able to duck the issue for long. More cases are lining up to force it to decide on the law of the land.

When people use numbers to make policy decisions, we should hold them to the same principles we hold judges to. We expect judges to carry out justice when they settle conflicts. We expect them to take responsibility for their decisions and to be accountable to the people they serve. We expect them to listen to the people affected by their decisions and give them a chance to explain why they deserve to win. Once judges decide, we expect them to justify their decisions by revealing their thought process in writing. We insist that people be able to appeal judges' decisions, either by going to a higher court or by bringing new cases that show new ways of thinking about the issue. We should expect no less of number crunchers.

"Treat likes alike" could be a simple formula for how judges decide. Cases that are alike should be decided by the same rules. Every child knows this formula for fairness, too, not in words but in feelings. Kids are keenly aware of who gets what and who has more or less. Early on, they think fairness means everyone gets exactly the same toys or the same-size piece of cake. By age 9 or 10, though, most kids come to understand that people are different and that sometimes it's fair for one person to get a bigger slice of cake.

This dining-table squabble is the classic kids' version of distributive justice—Who gets how much? The girl claims, "I'm older than you and growing faster. I need more calories, so it's fair that I get more cake." When she grows up, she might argue, "To each according to her needs." Her little brother thinks, "Our parents love us equally. I matter in this world just as much as you do, so I should get as much cake as you." When he grows up, he might argue, "We differ in a lot of ways, but we have equal moral worth." Both children think fairness has something to do with quantity. Both believe likes should be treated alike. They disagree on the standard for judging likeness.

I'm suspicious of Big Sis's motives. Maybe she does need more calories than her brother, but if so, I'd like to know whether Mom packs her a bigger lunch and stocks her soccer bag with energy bars. I'd like to know who baked the cake and why. Maybe Dad baked the cake to show his love for both kids and told them to share it. Either way, numbers alone can't settle questions of justice.

A division feels fair if everyone gets the right size piece. What's right, though? Ideally, people settle things by talking. They agree on a standard of fairness *before* they measure out pieces and cut the cake. Unfortunately, the real world operates more like sibling rivalry. The powerful cut first and talk later.

For the moment, let's pretend we live according to our ideals. We'll decide on a principle for fairness first. Is it right for cake to be distributed according to need? Or maybe it should be distributed according to merit—say, who gets the best school grades or who does the most family chores? Once we agree on a principle (assuming we can), we have to agree on how to apply it to real life. Suppose we agree food should be distributed according to need. If one sibling had a big lunch and the other didn't get any lunch, does the one who missed lunch need more cake? Does the reason for missing lunch matter—"I was too busy playing soccer to stop for lunch," or "I couldn't afford the cafeteria price"? If Big Sis is a growing teenager and does need more fuel than her 6-year-old brother, should she get to choose her favorite food to get the extra calories and stick her brother with the leftover broccoli?

The kids' dispute reveals the same problem at the heart of both justice and counting: red fish, blue fish, fast fish, slow fish—not one of them is like another. Likeness is in the eye of the beholder. Before we can treat likes alike, we have to decide which likeness is the most important one to equalize. Should we treat fish according to their speed and ignore their color? Or vice versa? For our cake dividing problem, should we treat kids according to their nutrition needs, their achievements in school, or their work doing household chores? For any problem of fair distribution, we have to choose a principle, something vague like need, merit, or effort. Once we've decided on a principle, we look for good ways to apply it so that likes are treated alike. This is one of the things that keep parents (and philosophers) awake at night. If they agree cake should be a reward for doing chores, should they expect the little guy to do exactly the same chores as Big Sis, or should they give her tougher chores to match her strength?

These are ethical questions. Hopefully, we answer them by talking, not by force. Even with lots of talking, we rarely reach answers. If we're lucky, we come to a temporary agreement on a principle of fairness. Measurement and numbers should come into play only after we've done a lot of talking and arrived at a comfortable resting place on the bigger principles. But we shouldn't rest in that place forever and put our measuring sticks on autopilot. The kids will keep on growing and changing and challenging us, as well they should. Like any good parent, as things change we should revisit our principles and how we put them into practice.

Wouldn't it be nice if scientists could come up with an objective way to count? Humans can be notoriously self-interested. Give them the power to decide other people's fates and watch them play favorites. Watch them be arbitrary and ornery. Watch them act out their grudges, tribal loyalties, and prejudices, all under the guise of law. No one's life pathways should hinge on such whim. If only we could find objective ways of counting, everyone would be treated equally. Life would be perfectly fair.

Sometimes I imagine this is what mankind wished for when the Fairness Genie appeared. We asked for objective rules to protect us from malign human discretion. Sure enough, the genie gave us point systems. In my fantasy, that's why numbers now make so many selection decisions: who receives public aid, whose insurance claims get paid, who gets hired, who gets admitted to colleges and universities, who gets bank loans, who gets scrutinized as a potential child abuser, who gets released without bail or let out on parole, and who gets promotions and pay raises. And that's why most large universities have taken to evaluating professors with point systems.

Here's a quick sketch (definitely not a photo) of how point systems work. Administrators assign points to the different things professors do—this many points for publishing an article in a prestigious journal (oh yes, even journals get points), fewer points for publishing in a low-ranked journal, and fewer points still for publishing a book. (I suppose the idea is that professors are notoriously long-winded; best not to encourage them.) Points for having their article cited in other people's footnotes. At the end of the day, each professor gets a number that represents his or her productivity and worth to the university. The point system treats every good professor the same, *if*, that is, the point system is a good measure of "good." It counts published articles but doesn't even try to measure their quality. Two professors might publish in the same journal but one's path-breaking idea counts the same as the other's rehash of an idea he's published 16 times. Last time I checked, no points were awarded for giving a colleague helpful feedback that helps her get an article published. No points for brainstorming paper topics with students or for helping them through emotional crises. The genie apparently hasn't figured out how to measure helpfulness, but it has given us what we wished for.

Or has it? We wish for firm rules so that other people can't decide our fate on a whim, or worse. Point systems don't have human faces, but people lurk inside every point. The people who design the systems may be invisible but they're human and they have their quirks and biases just like the rest of us. They think, to judge from some predictive algorithms, that black prison inmates released on parole are more likely to commit crimes than white parolees. They think, to judge from American education policy, that teachers who teach students testable facts are better than ones who teach students how to explore and learn on their own. They think, to judge from university point systems, that a jargon-filled article only specialists can decipher is worth more than a lecture that explains a complicated subject to students.

The Fairness Genie is a creature of our own wishful thinking, and it has been with us for a long time. Centuries ago, farmers knew that their masters exploited them by tinkering with measuring instruments. They wished for "one law, one measure" and dearly hoped that uniform measures would protect them. Standardized measuring instruments do make it harder to cheat and play favorites, but they can't prevent the strong from lording it over the weak. While the Fairness Genie is looking the other way, humans build their biases and their notions of good and bad into their counting methods.

The same people who once lorded it over employees (or criminal suspects or poor people) can now hide their biases in algorithms and point systems. Most of the algorithms used by large organizations today are designed and owned by commercial firms. The owners swear by the objectivity of their tools, but they won't let anybody peek inside to see exactly how the tools work. If you think an algorithm has treated you unfairly, you can't challenge it because you can't learn what information about you it has and how it uses the information. You can't learn because the inner workings of commercial software are trade secrets and protected by law.

What's to be done? We should start by sending the Fairness Genie back to the fairy tales where it belongs. Then we should demand—and our judges should grant—the right to cross-examine any counting systems that have the power to penalize us.

Even young children know that people pull the counting strings. I hadn't seen my neighbor's grandchild for a long time. "How old are you now, Konnor?" I asked him.

"Nine."

Grammy Tami looked at me. "His birthday is on Sunday," she said.

Konnor dropped his eyes to the table. "My birthday's in 2 days. I'll be 9 on my birthday."

Another pause. Konnor looked at his grandmother. As if feeling compelled to justify a white lie, he explained: "Three weeks ago, I asked my mom if I could say I'm 9. She said I can."

Grammy Tami rose to the occasion. "Yes, you can, Konnor."

At age not-quite-9, Konnor already has an inkling that the way adults count age violates the principle of "treat likes alike." He knew—and got his mom to agree—that there's no meaningful difference between 9 years, and 8 years, 49 weeks. But as he grows up, Konnor will find that some of the small differences he and his mom think don't much matter count for a lot.

If Konnor dreams of playing baseball in the major leagues, he'll have a leg up. His birthday is August 27. Kids get sorted into youth teams according to their age. The cutoff date for American youth leagues is July 31. The team for 9-year-olds includes kids who will turn 9 anywhere between August 1 and July 31. Konnor will be one of the oldest kids on the team. He'll play against kids who could be anywhere from a month to 11 months younger. And at age 9, as any parent knows, 11 months makes a big difference in physical development. More major league players in the U.S. have birthdays in August than in any other month. If Konnor lived in Canada and played hockey, he'd be at a disadvantage because the cutoff date for Canadian youth hockey is January 1. He'd be one of the smallest kids, unlikely to move up in the hockey world.

Assigning kids to teams by age doesn't seem like fair competition when height, weight, muscle mass, and agility matter more for athletic abilities than the calendar. Matching kids by their playing experience or coordination might be fairer, but these things are harder to measure. And changing a measuring stick to be more fair is easier said than done. There are always people who come out ahead with the existing rules. They tend to believe the current measuring stick is fair and resist change. It's not only the people being measured who resist change. People who've been using a measuring stick for a long time hate to give up a tool they know how to use.

Many elite colleges and universities in the U.S. rely heavily on the SAT (Scholastic Aptitude Test) to decide which students to admit. The test comes in for lots of criticism because students can boost their scores by taking intensive—and expensive—test preparation courses. Students whose parents are highly educated and who are lucky enough to attend good schools have a leg up on the test. The SAT seems to open doors for students who are already privileged and close doors for those who've already been left behind.

The company that runs the SATs recently decided to measure some of the circumstances that can help or hinder student learning before they take the test. The company reports these so-called adversity scores to colleges, along with each student's

SAT scores. Predictably, the new measuring stick stirs up controversy. Some highly educated and wealthy parents object to it because it changes the rules of the college admission game they've learned to play so well. Some critics think the adversity score doesn't do enough to help disadvantaged students, because it doesn't measure each individual student's circumstances. Instead, it uses information about neighborhoods, such as per-student school spending, and pins the same information on every student in a neighborhood. The kid whose family barely gets by in an upscale neighborhood doesn't get a score that reflects his struggles. The scoring system ignores many hardships that can hinder a student's learning. Parents with high income and advanced education might also have disabilities, divorces, or drug addictions that complicate their children's lives. Students with dark skin in well-to-do neighborhoods with good schools might suffer the crippling stresses of subtle prejudice. Details matter. No measuring stick can capture all the factors relevant to fair competition.

Many college admissions counselors welcome the new adversity score, however flawed, because it lightens some of the emotional burdens of playing judge. They want to be fair and they understand the problems with the SAT, so they've been mentally adjusting test scores up or down to account for students' backgrounds. Now they have a standardized measure, one handed to them by someone else. (Thank you, Fairness Genie.) The measure feels less subjective than their intuitive adjustments because someone else wielded the measuring stick, and because the adversity score applies the same criteria to everyone. Still, the new score lifts only some of the burdens of playing judge. Admissions officers know it's not perfect, so they'll continue to make intuitive decisions. They'll still use their own recipes for combining test scores, student essays, interviews, and the school's vision for its student pool. Most admissions officers probably welcome that freedom. If they wanted a job cranking formulas all day long, they wouldn't work in a college admissions office.

No measuring stick can take into account all the contextual factors that matter for fairness. Sometimes we don't know what matters for fairness. Which "likes" should be likened? Age or height for young athletes? For public assistance, should it be need for money or willingness to work or do community service? Our ideas about what matters are constantly changing. Differences that seem important at one time come to seem unimportant in another time. Sometimes we have a good idea of what matters but don't know how to measure it. A less-than-perfect measuring stick is better than none, so long as it always remains open to challenges and revision. No measuring stick should be forever. "It's not very good but it's the best we have" shouldn't be an excuse for doing nothing. When the measuring stick doesn't measure what matters, figure out why and change it.

Depending on the laws where Konnor lives, one year, at the stroke of midnight on August 26, he'll suddenly become legally able to drink. His question about counting his age will have even more urgency when 3 weeks could make the difference between getting arrested and not. The stakes will be higher. His mom's permission to fudge won't do him much good while he's talking to a cop. But by age 9, Konnor has already imbibed an important political lesson: you may not be able to change the metric, but you *can* argue with authorities about counting that feels unjust. You might be able to talk a cop into giving you a wink and a warning.

Adults understand that cutoff points between small increments can make a huge difference. A few dollars more in income can cost you eligibility for income assistance, subsidized health insurance, or scholarships. A tenth of a point on your grade point average can keep you out of a college or cost you your financial aid. At times like these, we regret ever asking the Fairness Genie to take the people out of counting. We wish for a little compassionate human intervention into rule-by-numbers. We hope that someone with power over us won't act like an automaton but will instead think, "Let me look into this. Perhaps I can bend the rules here."

If Konnor gets a job driving for Uber, he won't have a human boss. He'll be bossed by numbers. (That's if human drivers aren't obsolete by the time he's old enough to drive.) Uber's algorithms will hire him, fire him, supervise him, evaluate him, and pay him. The algorithms will count the times he brakes smoothly and the times he brakes harshly. They'll detect when his phone moves and warn him that he should have his phone mounted, not in his hands. The algorithms will count Konnor's hours on the job, which for Uber means the hours he's online with the Uber app. To prevent him from discriminating, the algorithms will count the times he declines to pick up passengers who ask to be dropped off in low-income neighborhoods. If his "ride acceptance rate" falls below 80 or 90 percent, the algorithm will fire him. If his customer ratings fall below a 4.6 on the 1-to-5-star scale, he'll be fired.

All this instruction and monitoring will happen without Konnor ever being able to talk to a human. If he gets fired, he won't hear "We're sorry we have to let you go." He'll learn that he was fired when he tries to log in and the app won't let him. If any passengers bully him, proposition him, or harass him with hate slurs, the algorithm won't know. He can try to get help from his "community support representative," but the rep isn't a person; it's an email address at an email equivalent of a call center. He'll get a robotic answer that probably won't have anything to do with his problem.

Driving passengers requires good driving skills and good people skills. Algorithms are pretty good at measuring people's relationships to their machines. They can count smooth braking and hours online. Algorithms aren't so good at managing human relationships. They can't recognize when a driver deftly handles an obnoxious rider or calms an anxious one. If a passenger gives a driver a low rating because she doesn't like "his kind," the algorithm will never know. Algorithms can't detect shared stories, moments of human connection, or the driver who goes out of his way to help a passenger. People skills and difficulty of the job don't count in Uber's evaluation of its drivers.

Before there was Uber, was it any better to work for a cab company? Am I romanticizing human bosses? Alex Rosenblat listened to stories of Uber drivers in 25 cities. My portrayal of Uber's management-by-algorithm comes from her book,

Uberland. Most of the drivers told her they loved the freedom to choose their own hours and be their own boss. But as one driver put it, "It's better [not having a boss] except when something goes wrong."

In some areas, automated judgment technologies do deliver good results. Algorithms guide machine tools to turn out airplane parts down to an nth of a micron, and planes are safer for it. Automatic pill counters save pharmacists from tedious work and probably make fewer mistakes. But in areas where human relationships are important, we ought to think hard about what's lost when one side of the relationship is no longer human.

I feel sorry for today's students who write essays for an audience of machines. Automated essay scoring is now a minor industry in the education world and several states have adopted it for their standardized testing. Automated scoring saves teachers time and school systems money. Students get their results more quickly. Companies that sell these systems bill them as Fairness Genies: a student's grade is no longer tied to one teacher's tastes. As you know by now, though, *somebody's* taste lives inside those automatic essay graders and tells them how to grade. Now it's a committee of teachers that decides what makes a good essay.

Automated essay scorers check for correct grammar, spelling, word usage, sentence structure, and organization. Does the essay have an introduction with a statement of purpose? Does it provide supporting evidence for its arguments? Does it state a conclusion? Too bad for the student who wants to be a poet or novelist. An auto-grader knows how to evaluate the nuts and bolts of writing, but writing is about communicating, not only technique. When I write, I conjure up my readers in my mind and hold imaginary conversations with you. I imagine you laughing at my jokes, arguing with me, having a light bulb go off in your head, and sometimes telling me I'm clear as mud. If it weren't for you, I wouldn't know how to get my ideas across. If I didn't think you were going to read me, I'd feel lonely and useless sitting solo in my study. Good writing moves readers, makes them question, makes them see things in new light, makes them laugh, stirs them to outrage, shows them beauty, sends them exploring, changes them in some way. Automated graders can't respond to writing in any of those ways. They can't enjoy word play, get a joke, feel surprised, or appreciate creativity. They will never be able to get into a relationship with a writer.

Anyone who's ever tried internet dating knows that you don't want to get into a relationship with everyone. In some contexts, though, you might prefer being in a relationship with an unappealing and possibly biased human to existing as a data bit inside a computer. As one client of social services put it, "You can teach people how you want to be treated. They [caseworkers] come with their own opinions but sometimes you can change their opinion. There's opportunity to fix it with a person. You can't fix that number." From the perspective of caseworkers, automated decision making deprives them of their most important tool—hope. "People who have gone through a trauma want some hope that it's going to get better. That somebody's paying attention, that they're not in this alone. That's what I think we did [before automation]. We listened to what they had to say and acted on it so that things could get better."

Giving people an "opportunity to fix it" with another person empowers them. Without a chance to exercise some control over what happens to us, we become helpless rats in a cage. We acquire a sense of power by being able to make a difference in how other people treat us. Okay, some people get a sense of power from winning video games and coaxing smartphones to do their bidding, but these devices teach them to stare at their LED screens like zombies, oblivious to anyone else. Only interacting with humans can teach the social and civic skills that Konnor learned when he questioned his mother and grandmother about counting his age.

Opportunities to challenge decisions that harm you are the essence of what we mean by fair procedure. There's no bargaining with an algorithm, no questioning, no right to see the evidence that knocked you down. If these automated decisions were trials, you'd have a right to see the evidence against you. You'd have a right to question the prosecution and make your case to a judge or a jury. If an algorithm's adverse judgments about you were criminal convictions, you'd have the right to appeal them to a different judge. You'd be able to present yourself to another human being, show them how the world looks from your shoes, justify yourself, and try to win their understanding and sympathy.

These same principles of law ought to apply to counting tools that alter people's life pathways. No one should be at the mercy of an algorithm. When numbers can't do justice, there should be a way for people to have recourse to humans. In some areas, such as child welfare and criminal justice, there are provisions for humans to look at unusual cases and override the computer. Most of the algorithms used in social services, criminal justice, and education come with stern instructions from their developers: "This algorithm is not to be used to make decisions. It is intended to be used only as an aid to your decision making." Such commands have about as much force as parents' words to teenagers heading out the door.

Who makes a better judge: a quirky human or a staid calculating machine? The question updates an old chestnut in legal philosophy: Which is more just—hard-and-fast rules that apply to everyone or flexible rules that allow for some human discretion? On the one hand, justice means a government of laws, not men. Clear, firm rules should govern everybody, from the weakest to the most powerful. On the other hand, we want the people who have power over us to see us fully and take our unique circumstances into account. When adhering steadfastly to a rule would be cruel, justice must make room for exceptions.

Realistically, people who use algorithms and point systems to judge other people aren't going to be easily swayed by moral argument. Often they adopt automated decision tools precisely to disrupt human bonds. Politicians aiming to trim welfare programs *intend* to prevent caseworkers from empathizing with clients and bending rules to help them. Still, it's worth fighting for justice in counting.

Sometimes frankly subjective measures are better than ones that might seem more objective. To decide what belongs in the category we're counting, we select the most important features of things, but important to me isn't the same as important to you. The only way to find out what's important to people is to ask them. Empower them to measure themselves and what they care about, or at least empower them to help you measure what they care about.

Do peace building efforts in war-torn countries work? Outside experts usually measure things their theories tell them should promote peace. They count how many government officials are chosen by election, for example, because they believe democracy promotes peace. They count how many police and security guards are on the payroll, because they assume more guardians mean more peace. They measure their own assumptions instead of what really matters—whether people feel like they live in peace. The Everyday Peace Indicators project holds focus groups to learn exactly how people experience peace or lack of it. Group leaders ask, "What are the signs that tell you whether or not there is peace in your village?" In Afghanistan, the answers are disarmingly mundane and heartrending.

"We see girls going to school." (Unspoken: that means the Taliban isn't banning girls from education or closing schools.)

"We go to the hospital in Lower Pachir during the night" (because we don't dare go in daylight).

"We see Afghan Local Police outposts." (We feel protected, knowing they're there.)

"We see our village leaders going to the district governor's office." (Local government must be functioning.)

"We see antennas on houses." (The Taliban often bans TV, radio, and satellite, so antennas are a sign that they don't control our village.)

"The shops in our village have never been broken into by thieves."

"ISIS confiscated the cattle of people of Pachir village."

"We see employees of mobile communications companies bringing fuel to the towers in our village."

"We see many people come to our district for weeding opium." (In the U.S. or Mexico, opium cultivation would no doubt be a sign of trouble. For Afghans, it means economic prosperity, a necessary ingredient for peace.)

These statements are ground-level indicators. They tell whether people can go about their daily lives. To people who've known decades of war, being able to go about your daily life is what peace means. Daily life isn't the same everywhere, so the indicators differ from place to place. In some black South African communities, toilets are usually outdoors and shared by many households. People in those communities said that being able to urinate outside at night meant they felt safe. Precisely because the Everyday Peace project doesn't use the same

measure in every place, its indicators are meaningful to the people whose lives they try to measure.

To convert these indicators into a measuring stick, researchers turn the statements into survey questions with 5 possible answers. For example: "How often do you see girls going to school? Never? Rarely? Sometimes? Often? Almost always?" The researchers give each answer a number from 1 to 5, with 1 for the lowest, least peaceful answer, and 5 for the highest. Add up all the answers, find the average, and we have a way to measure trends in peace over time or to compare levels of peace in different regions.

It's tempting to think that school attendance numbers would be a more objective measure than statements like "We see girls going to school." Alas, attendance records aren't always trustworthy. Teachers might lower the numbers so as not to arouse the Taliban, or they might raise the numbers to encourage more girls to come to school. *How many* girls are in school isn't so important as the fact that *any* girls feel safe walking to school or sitting in outdoor classes. Counting police on the local government payroll might seem like a more objective way to measure security than statements such as "We see police at their outposts." But payroll tallies wouldn't tell us nearly as much as villagers' words. When villagers say seeing police makes them feel safe, we know that police are showing up for work and that when they do, they're protecting villagers, not raping, robbing, and torturing them. Numbers of cattle raids and shop thefts might seem like objective measures of failed peace-building, but those numbers wouldn't convey how a single incident can fill the atmosphere with terror. One shop theft might be all it takes to discourage people from opening shops. One cattle raid could make people in surrounding villages afraid to graze their animals. Here are cases where 1 amounts to more than 1. When people say they've noticed, we know something unpeaceful has penetrated their psyches.

Whenever we measure, we have to make trade-offs. We can have calculator-like precision if we stick to easily countable items such as police rosters. We can have cinematic realism if we muck about in mundane details of daily life, though we'll get a fuzzy montage, nothing you can put your finger on and report, "I've got 157." The movie is more likely to make people feel understood than the police roster. It's also a more accurate measure of peace if we think peace is a quality of daily life and a state of mind rather than a police payroll sheet.

The Everyday Peace Indicators capture people's sense of peace so well because the project leaders built the indicators from messy talk. They listened to stories. They asked questions. They sat through plenty of gossip and off-topic digressions, too. They distilled the messy talk into short phrases that could be listed, easily understood, posed as questions, and transformed into statistics. They simplified, but they didn't abstract so much that the indicators lost connection to lived experience.

Objectivity is supposed to be a universal point of view. For trying to understand peace, objectivity misses the point. Peace is peace to particular people, not to an imaginary omniscient observer. Not only does subjectivity provide a more accurate measure here. Developing measuring sticks from the bottom up empowers people by giving them a microphone to say what matters to them. If we're going to use statistics to design policy, the people being measured and the people whose lives will be affected by the numbers should have a voice in determining what gets counted. This is basic democracy.

Ethics, in a nutshell, comes down to the question "What's the right way to act?" In one of the most famous puzzles about ethics, you must decide what to do when 5 people's lives are pitted against 1 person's life. The 5 have somehow been tied to a trolley track. A train is screaming toward them. The tracks divide before the train will run over them. You happen to be standing beside the track next to a switch that will divert the train to another spur where there's 1 person tied to the tracks. Should you throw the switch? If you do, you would make the train kill 1 person instead of 5. Would you be a hero for saving the 5 or a murderer for killing the 1? What's the right thing to do?

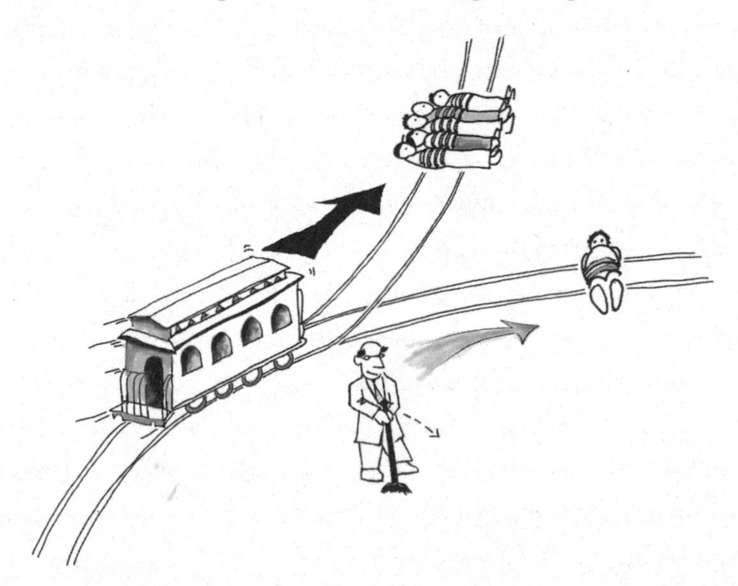

The trolley problem is a simplified version of many policy problems where there's an unavoidable trade-off between harming a few people and benefiting many. With a little tweaking, the trolley problem could become a morality play about economic development. Five farm families, tied to their land, are about to be financially and socially ruined by a shopping mall that will create 300 jobs. Five blocks of low-income apartments are about to be sacrificed for a convention center that will revitalize the downtown. There's someone at the switch in these situations, but it's not a random bystander. It's the town zoning board or the mayor, or maybe a judge who decides whether the local government can take the farms and homes by eminent domain.

Those are tough problems, loaded with political overtones, so let's go back to the trolley problem. Some people think, "Of course you should throw the switch. Maybe hesitate long enough to feel pity for the 1 man, but remember, *you* won't kill him, the train will." Other people think, "It's wrong to kill people so you shouldn't take an action that you know will kill someone." These answers represent the two major schools of thought in moral philosophy.

One tradition has a formula for solving all moral problems: do whatever will yield "the greatest good for the greatest number." If you can save 100 people by killing 1 person, killing the 1 is the right thing to do. This is the tradition associated with Jeremy Bentham and utilitarianism. It can be hard to predict what all the good and bad results of an action will be, but once we've identified them, doing the math is easy. The numbers will tell us the right thing to do.

The other tradition asks us to reason our way to right answers. We're supposed to find general principles for determining what's right and wrong. Do unto others as you would have them do unto you. Before you do something that you believe won't cause any harm, ask yourself, "What if everyone did it?" This is the tradition associated with Immanuel Kant and deontologists. It's complicated to find principles everyone agrees on, but once we do, applying them to real situations should be easy. Except it never is, and that's what keeps philosophers employed.

Philosophers have come up with many variations on the trolley problem to help us tease out our assumptions as we think about morality. My favorite version is by Judith Jarvis Thomson and it goes like this: Five people have been tied to a railroad track. You happen to be standing on a footbridge over the track while a train screams toward them. You're the protagonist who has to figure out what to do. In one of those weird details that only novelists and philosophers can conjure up, there's an obese man on the bridge with you and he's leaning far over the railing.

Let's pause here for a moment. What do you think will happen next? If I were writing this novel, I'd have the protagonist talk to the man. I'd have her say, "Do you see what I see? Is there anything we can do? Who tied them to the track? Let's remember every detail so we can provide evidence to the police. What do you think is going through their heads right now? If you were down there, what would your last thought be? Shall we say something to them—maybe yell that we're sorry, we'll remember them, tell their story, and try to help their families find justice?"

If I were writing this story, in a million years I wouldn't imagine that the protagonist would think to herself, "I could push this man over the railing so he'd fall onto the tracks. His body mass is so big that it would stop the train. By killing this man next to me, I could be a hero and save those 5 people." But that's exactly the thought Thomson put into the protagonist's head.

And that's exactly how utilitarian philosophers have asked us to think about whether it's wrong to kill someone. We aren't supposed to get hung up on whether it's *always wrong* to kill someone, and why or why not. We're supposed to decide whether it's *sometimes right* to kill someone and we're supposed to decide on the basis of the numbers. 5 against 1. Majority wins. Like point systems for evaluating professors, utilitarianism purports to give us an objective way out of moral quandaries.

The other versions of the trolley problem all feature many people tied to a train track and a bystander who could save them by doing something that will kill fewer people. All versions stack the deck by pitting the interests of the many against the interests of the few. In a democratic culture, the go-to solution counsels killing the few to save the many. Yet, there's a strain in democratic culture that wants to protect minorities from tyranny of the majority, to fight mobs with reason. That's the strain that inspires the trolley problems.

Judith Jarvis Thomson made up the obese-man story to refute moral reasoning by arithmetic and to make us question utilitarianism. She argued that the man has a right to his life, a right that others can't take away no matter how benevolent their motives. Many philosophers in Kant's tradition assert in one way or another that it's just plain wrong to kill or harm someone deliberately. It's wrong even if you believe that you're doing it for a larger purpose or that you're not the one physically striking the blow.

Not surprisingly, when people ponder what's the right thing to do in the trolley problems, they're highly susceptible to the power of suggestion. Changing a few details in the wording of a moral dilemma can change how people think about it. One researcher created an experiment to compare how people think about two descriptions of the trolley story. In one version, Mark, the protagonist, hates the man whose life could be sacrificed to save 5 people. Mark thinks to himself, "I don't give a damn about those 5 men but this is my chance to kill the bastard." In the other version, Walter, the protagonist, faces the same choice—do something to kill 1 man in order to save 5—but we don't know anything more about him except his name. Most subjects in the study thought it would be right for Walter to kill the lone man but not for Mark. They (and many philosophers) think that when we're deciding what's the right way to act, numbers aren't everything. Motives matter.

Math doesn't work as a solution to moral dilemmas because people don't think about themselves in the abstract. We think of our lives as stories, not arithmetic problems. We think of people as *having* character, not as being characters out of central casting. The math solution to moral dilemmas lets us duck questions of responsibility. How did 5 people get tied to the train tracks in the first place? Not by a mythical god or an act of nature.

Framing problems as accidents waiting to happen distracts us from searching out human causes. Somewhere along the way to an accident, humans made choices that led to it. What were those choices? Did the people who made them impose them by brute force, as the trolley scenario suggests? When we're faced with an accident waiting to happen, of course we need to decide how to act in the moment. But if, instead of reading the trolley story as an accident waiting to happen, we read it as a contest between 1 strong thug and 6 of his weaker neighbors, we'll imagine some different ways of responding other than sacrificing yet another innocent victim.

There's yet more danger in using arithmetic to solve social problems. Imagine you go to the hospital to visit your sick brother. While you're there, a scanner reveals that all your organs are perfect matches for 5 patients waiting for transplants. Without receiving an organ in the next few days, each of them will die. If hospital doctors follow the simple formula "Greatest good for the greatest number," they should kill you (humanely) and harvest your organs so that 5 people may live. If doctors did what utilitarianism says is morally right, no one could trust them anymore. People might avoid hospitals unless they're in such agonizing pain that dying peacefully sounds appealing. If judges upheld killing an innocent person to benefit several others, as in the trolley problems, no one could trust the legal system. Here's where Kant helps. If everyone acted on utilitarian principles, social trust and community bonds would wither.

"The greatest good for the greatest number" is intuitively appealing. Many problems don't have a solution where everyone can come out ahead. In those cases, some people have to make sacrifices for the greater good. When we make those decisions, we do so with regret. We say, "We're sorry and we're very grateful for your sacrifice." Ideally, we compensate them in some way, too. "The greatest good for the greatest number" works as a simple slogan to describe these kinds of cases. But not every case is this kind of trade-off. Before we resort to arithmetic to solve a social problem, we should ask two questions. Can we deal with this problem by finding its cause and preventing more problems like it? And, what would life in our community be like if everyone decided to act on their interpretation of the simple slogan?

Utilitarian morality isn't just for philosophers. No one can get a degree in business, law, medicine, or public policy without learning how to think like a utilitarian. But instead of wrestling with far-fetched trolley problems, people in these professions learn to solve real-world problems with cost-benefit analysis, or cost-benefit for short. Cost-benefit, as the name implies, lets us compare the costs and benefits of different programs so we can choose the one that will give us the most bang for the buck. If we're thinking about investing in an activity, cost-benefit tells us whether we'll get back more than we put in—maybe not in dollars, but in other valuable things like health and safety.

You may not think utilitarian morality figures into your personal life but when you struggle with a hard decision by making a list of pros and cons, you're doing a rough cost-benefit analysis. The difference between what you do and what the high-powered policy analysts do is that they put numbers on everything. You might be content with plus and minus signs or smileys and frown faces. Policy analysts want numbers. Not just any numbers, but money prices. And now the federal government wants money prices, too.

Thanks to Presidents Clinton and Obama, whenever government agencies propose a new regulation, they have to put it through a cost-benefit analysis. If the Environmental Protection Agency wants to regulate auto emissions, it must first predict all the good and bad consequences of the regulation. Then it has to put a dollar value on every good and bad result. The agency won't be allowed to issue the regulation unless the benefits outweigh the costs. Granted, it's hard to measure all the benefits and costs of a program and it's even harder to put a monetary value on some of them, such as human life and health. But in a world of limited resources, champions of cost-benefit say, it's better to have some way of estimating how best to use them than no method at all.

You could go to a textbook to learn how to do this stuff, but an old nursery rhyme teaches it better.

For the want of a nail the shoe was lost,
For the want of a shoe the horse was lost,
For the want of a horse the rider was lost,
For the want of a rider the battle was lost,
For the want of a battle the kingdom was lost,
And all for the want of a horseshoe-nail.

You might know this ditty as a children's lesson about diligence. Take care in your work. Don't stint on even one small thing or you might cause disaster. For the would-be policymaker, the ditty holds another lesson: every action we take—or don't—has consequences that ripple out into infinity. Got it? Now let's take this horse on a wild ride through automobile emission standards.

If government restricts auto emissions, fewer children will develop asthma. Their medical costs will go down. Their parents will take fewer sick days and will be more productive. The parents' work will enable other people to do and buy more things. The extra productivity will create more jobs. We'll need to count all those things and put a value on them. And asthma's only one of many things that will be affected by lower emissions. There'll be less acid rain and fewer greenhouse gases. Agriculture will be more productive. Forests will generate more timber. Global warming might proceed a tad more slowly. If it does, droughts, forest fires, hurricanes, and floods will be less severe. All for a bit of emission control.

Having galloped through some benefits of tighter emission limits, let's turn around and ride through some costs. Car manufacturers will have to invest in new technology. They'll take a hit to their bottom line. To compensate, they might hold down wages or raise the prices of their cars. Employees and customers will have less money to spend. Maybe they'll cut back on beer. Pub owners and breweries will lose money. Maybe people will cut back on day care. The kids deprived of day care won't grow up to be as clever as they might have been. They won't earn as much or contribute as much in taxes. All because of an emission regulation.

I've let my imagination run wild because that's my point. Cost-benefit analysis is an exercise in imagination before it becomes an arithmetic problem. Every action causes ripples in the universe. Policies and programs have no natural stopping place for their ripples, no barrier like the shore of a pond. A cost-benefit analyst has to set an arbitrary point where he'll stop counting ripple effects. Should he stop at the horse? The rider? The battle? Should he count what happens after the kingdom is lost? Maybe the people are better off under the new king. Advocates for and against a program can manipulate the ratio of costs to benefits simply by imagining more ripple effects on the good side or the bad side.

Even if researchers don't deliberately manipulate their results, cost-benefit analysis is ethically troubling. For scientists, honesty is the first ethical imperative: Don't lie about your data. Don't fudge it. Don't hide any of it. And for heaven's sake, don't make it up. Cost-benefit analysts can't possibly be honest in these ways. The method relies on anticipating what will happen after a policy goes into effect—in other words, on making up the future. The method relies on hiding data, too. Analysts can't possibly count all the ripple effects they know or can guess might happen. Cost-benefit studies are always speculative, incomplete, and arbitrary. For all their jargon and math trappings, they're anything but the rigorous science they pretend to be.

Like the trolley scenarios, cost-benefit analysis frames issues as arithmetic problems. Which adds up to a bigger number—the costs or the benefits? And like the trolley stories, cost-benefit analysis raises other ethical issues. By framing policy decisions in terms of numbers, it pushes aside moral reflection and evades questions about moral responsibility.

EXHIBIT A: Smoking and Vaping

Smoking nicotine can lead to lung cancer, heart disease, and many other diseases. The Food and Drug Administration (FDA) now has authority to regulate e-cigarettes as tobacco products. Should it ban vaping or restrict it to adults? In the 1990s, several cost-benefit studies concluded that the benefits of cigarette smoking outweigh its costs. Are you ready for this? The more people who die early, the less money society will have to spend on their medical care, pensions, and housing. Apply that same reasoning to vaping and the benefits are even greater because teenagers are the main users of e-cigarettes. We don't yet understand all the health effects of vaping, but we can guess that the sooner people start vaping, the sooner they're likely to suffer from vaping-related diseases and die. Think of all the money we'll save on taking care of them in old age. If governments and insurance companies want to save money, they should *encourage* smoking and vaping. By this cynical logic, government shouldn't try to stem the opioid crisis, either.

The smoking studies were exceptionally perverse and gave cost-benefit analysis a black eye. Few studies today are so brazenly callous. No one advocates analyzing vaping the same way. But even if the early smoking studies were rare bad apples, they reveal something rotten in the method. Cost-benefit analysis has no moral guidelines about what things to count. Without any moral constraints, analysts can count deaths as benefits to society because once people are dead, it doesn't cost anything to keep them alive. Letting people slowly kill themselves with dangerous products looks like the right thing to do.

EXHIBIT B: Prison Rape

In 2003, Congress passed the Prison Rape Elimination Act. It took nine years for the Department of Justice to issue rules about how states should implement the law. Before issuing the rules, officials performed a mind-boggling 168-page cost-benefit analysis. They guessed what the program would cost ($470 million). Then they made a list of 17 different types of rape and guessed how much each kind was worth (between $300,000 and $900,000). (Hang on: I'll answer your question in a minute.) At the time of the study, there were about 160,000 rapes per year. If (pure speculation) the new program prevents only 1 percent of the rapes (1,600 rapes), and if (pure speculation) each prevented rape adds $500,000 worth of well-being to the prison population, the program would yield $800,000,000 in benefits, way more than it costs. The program gets a green light.

A moment ago, you asked why anyone would think of rape as being worth money. Rapes that don't happen are the benefit of a rape-prevention program. In cost-benefit analysis, a program's costs and benefits must be converted into dollars so that we're comparing apples to apples. The rape prices are meant to capture what it's worth to an inmate *not* to be raped. If you're wondering how anyone determines the value of not being raped, cost-benefit has two ways to find an answer. One: ask people what they're willing to pay to avoid being raped. Two: ask people how much money they would demand in return for letting someone rape them. (I know this sounds preposterous, but if you don't believe me, look up "willingness to pay" and "willingness to accept" in any textbook.) In other words, cost-benefit analysts run an imaginary protection racket to find out whether the Department of Justice should try to prevent prison rapes.

Aside from the stupefying notion of putting extortion prices on rape, cost-benefit analysis asks the wrong questions. We know for a fact that prison inmates are often raped. Imprisoning people is somewhat like tying them to a railroad track knowing they'll be maimed. In this case, government is both the thug who ties people to the track and the bystander who could do something to prevent the catastrophe. Cost-benefit analysts ponder how much money government and its taxpayers should demand as the price of not tying people to the tracks. Instead, we should be asking, "Who's the thug?" and "Who could do something to prevent this disaster?" Instead of asking "What's the most efficient way to spend money?" we should ask, "What's the right thing to do?"

Cost-benefit analysis puts a modern twist on James Madison's method for calculating the value of slaves. Madison looked to laws made by powerful white men for the purpose of keeping black people down. Cost-benefit looks to the status quo to value things, too. It puts a math gloss on giving powerful people the right to continue harming other people.

If numbers aren't the best way to resolve ethical and political dilemmas, where should we turn for answers? Ultimately, we put our faith in democracy to select wise and experienced leaders. But here's the rub. Democracy runs on numbers. As civics courses teach, counting votes is the best way to find out the public's desires and translate them into policies. Elections enable large groups of people to pursue common ends without beating up on each other or resorting to a dictator. To be sure, democracy runs on deliberation, too. Before and during elections, people can talk their differences through and reach consensus. But eventually, when the caffeine has worn off and the people have run out of energy, talking will cease and votes will be counted.

Elections are an ancient form of public opinion polling. Ballots are the questionnaires. Candidates are the possible "answers" that political parties put on the questionnaire. Voters answer the questions by checking or not checking candidates. Occasionally voters get to say yes or no to a specific policy proposal in a referendum, but most of the time, they get to vote only for people, not policies.

In theory, before voters go to the polls, they've listened to candidates' speeches and they know what policies each candidate will enact if elected. Take a moment to look into your own voter soul and tell me whether you believe the theory. But let's keep the faith for a moment. Through the rituals of voting—showing up at a polling station and casting ballots—citizens' ideas and hopes are transformed into countable bits. The final numbers offer a seemingly definitive answer to what the people want. A winner can proclaim, "I've got a mandate from the people," though the election gave him no such thing. Election results ignore why people voted the way they did. No matter how unequal everything else that happens before the polls open (hold on, we're coming to that), elections reassuringly symbolize that everyone's voice is heard and everyone counts equally.

For all their virtues, elections aren't really up to the challenge. It's impossible to count citizens' thoughts and wishes about what we want from our communal life. The path from a person's mind to her representatives in government to what government actually does makes a pinball machine look straight and narrow. First there are rules for turning citizens into eligible voters. In pinball, you don't get to play if you don't have quarters to start the machine. In elections, you don't get to vote if you don't meet the ID and registration requirements. The same people who are supposed to carry out the voters' wishes decide who's eligible to vote. Oops. One after another, voters are funneled down the pinball drain and removed from the game by the very people they're supposed to choose to lead them.

Next, there's the process of counting ballots. In pinball, players can tilt the machine so long as they do it just slightly enough to avoid getting caught. In elections, the players can tilt the outcome by discarding ballots, "losing" them, miscounting them, and hacking the voting machines, so long as they don't get caught, and sometimes if they do. Last, there are rules for converting votes into seats in legislative bodies. In pinball, the machine's designers decide how players can earn points. In elections, the rules for winning power—seats and offices—are made by people already in government or by people powerful enough to be in the room where election rules are written.

These are problems baked into democracy from the beginning and we haven't yet mentioned how donors and lobbyists, news media and social media, trolls and bots change the course of elections. Counting votes, the simple tally at the heart of democracy, isn't so simple as we once thought. Popular will might be the quintessential thing that can't be counted.

For all their defects, elections are still the best means we have for holding our officials to account. But those defects are substantial. That's why, before and after we go to the ballot box, we have to find better ways to count everything we do and care about.

When Abraham Lincoln dedicated the cemetery at Gettysburg, he might have started by toting up the battle casualties. By the standards of any age, 50,000 is a staggering number. Lincoln was humble enough to know that neither numbers nor words could do justice to the moment or to the events that led up to that moment. "We cannot hallow this ground," he said. "The brave men, living and dead, who struggled here, have consecrated it, far above our poor power to add or detract. The world will little note, nor long remember what we say here, but it can never forget what they did here."

Toni Morrison called on Lincoln in her acceptance lecture for the Nobel Prize in Literature. Even as she was being honored for her supreme prowess with words, she followed Lincoln's example to warn of language's limits. "Language can never 'pin down' slavery, genocide, war. Nor should it yearn for the arrogance to be able to do so."

And so it is with numbers. If I have one message about counting, it is: Stay humble. Numbers are the products of our poor power to make sense of our lives. They aren't truth meters. We shouldn't use them as arbiters of political conflicts or as answers to ethical quandaries.

We need humility when we count so we don't mistake numbers for life. Numbers grow from human imagination. We shouldn't put forth numbers or rely on them without examining the human judgments they stand on. We need humility when we count so that we don't go it alone, so that we ask the people whose lives will be affected by our numbers to help us count better.

Numbers contain the stories people tell each other and themselves. Hold your ear to a number as to a seashell and listen to its whispers.

Epilogue: Counting Goes Viral

What do you do when you've been exiled from the world you know? You count. You take it one day at a time. You tally your toilet paper. You keep tabs on your phone battery. You calculate how much longer you'll be able to pay your rent. You take your temperature. You glue yourself to the news and listen to the daily numbers as if to an oracle.

When the coronavirus first emerged in China, most everyone aside from public health specialists believed it would stay in China. Then cases of COVID-19, the disease caused by the virus, appeared in other countries, soon followed by a death or two. The illusion of national sovereignty evaporated. Closing borders failed to keep the virus out. Countries began running their own domestic tallies, while keeping an eye on the international scoreboard. In the viral Olympics that replaced the Tokyo games, nations seek to win by getting low scores.

Every number begins with a human decision about what to count. The scorekeepers started by counting new COVID-19 cases. Within days, though, people grasped how meaningless the reported numbers were. The daily case count can't tell us how many people are or have been infected. It tells us only how many people were tested and found to be positive at the moment they were tested. People who might be infected but have only mild symptoms aren't usually tested. They aren't counted. Some test results are false negatives. Those people aren't counted, either. Some of the people who test negative one week may become positive the next week. The tests take time to process, so the results give us a glimpse into the past, not the present. The experts and pundits explained all this, but they kept reporting the daily case numbers anyway and we kept listening for signs of hope.

Because the number of new cases so obviously depended on how much testing was done, the scorekeepers added a new Olympic event: testing. How many tests were countries doing per day? Per week? Per capita? How many test kits were available? How many were available in places that were hardest hit? In the U.S., people wanted to know why we had such low testing rates compared to China, South Korea, Italy, and other countries supposedly behind us in technological know-how. Testing is the main way to count people infected with the virus and it seemed the U.S. was hardly counting.

Low testing numbers generated a political whodunit. Journalists went hunting and identified some likely suspects. The

FDA had made it difficult for private labs to develop tests without going through a lengthy approval process. The CDC held a monopoly on producing test kits, but its initial batch of 90 kits was flawed. No one in the executive branch was coordinating the CDC, the FDA, state public health labs, and private hospital labs. And most intriguing, the CDC at first set narrow criteria for who could be tested—only people who'd recently been in China or who'd had close contact with a known case. With so few people eligible to be tested (and counted), there didn't appear to be an urgent need for more testing capacity. Here's a wonderful paradox about counting: If you decide a problem is small before you start to count, then you won't bother counting in a way that could show it's bigger than you thought.

As the virus proliferated, so, it seemed, did counting. How far can a cough project a lethal droplet? (6 feet, or maybe 20.) How many people are newly jobless? (22 million and still counting.) How many qualified FEMA personnel are available to lead field operations? (19.) How many bottles of hand sanitizer did one gouger stockpile before Amazon suspended him? (17,700.) How many fewer condoms would Kartex, the Malaysian contraceptive giant, be able to produce from mid-March to mid-April? (200 million.) How fast were virus-related cyberscams growing? (Complaints doubled in one week to 7,800.) And the most disturbing count of all, how many people have died from COVID-19?

COVID death counts are bedeviled by questions of how to count. Death numbers come into being by a kind of bucket brigade. If someone dies in a hospital, doctors record what they believe to be the primary cause of death and report it to a local or county medical examiner. If a patient hasn't had a virus test, then doctors have to make a judgment call about the cause of death. Older people are especially likely to have a multitude of problems that contribute to their deaths, but with COVID in the forefront of doctors' minds, they are more likely to blame it rather than other factors. If a person dies at home, usually a medical examiner declares a cause of death. Doctors and medical examiners report their numbers to a city or county health department, which passes the numbers up to the state health department. States pass their numbers to the CDC. Similar counting chains exist in other countries, with doctors, hospitals, and coroners passing numbers up to a national statistical agency.

The numbers at the end of a bucket brigade are only as good as the numbers at the beginning. Each person or agency in the brigade decides what to count as a COVID death. Initially the CDC counted only deaths in which a laboratory test had confirmed the virus was present. But early on, it was clear that doctors and medical examiners weren't systematically testing bodies or recording every deceased patient who had tested positive. How *could* they with skyrocketing deaths, makeshift morgues, and mass burials?

In New York City, for example, people who died at home weren't counted as COVID deaths unless they happened to have been tested and tested positive before they died, and their names had been reported to the city health department. The day after a reporter broke the New York City story, city health department officials announced they would start counting "probable" COVID deaths in addition to confirmed ones. Once officials decide to start counting people who probably had COVID when they died, the COVID death toll will rise without any change in the virus's behavior. Sure enough, less than a week after New York City started the new counting method, its death toll jumped by 3,700, increasing the city's number by about 50 percent and the U.S. death toll by 17 percent.

In the U.S. and around the world, some agencies now count people who probably died of COVID even if they hadn't had a positive lab test. "Probably" means that a doctor or medical examiner made a judgment call. That's not wrong or misleading. Every death certificate lists a primary cause of death based on a doctor's judgment. But numbers based on buckets full of judgments are hard to compare and easy to question.

Everyone wants to know how lethal the coronavirus is. Is it more deadly than the common flu? Is it more deadly in some places than others? The most common way to score its deadliness is to figure the number of COVID deaths as a percentage of the number people who tested positive. In Italy, it appeared that the death rate was between 11 and 13 percent. In the U.S., it was a little over 4 percent. In Germany, the death rate appeared to be around 1 percent. In China, it was down near 4 percent. Israel, at 0.35 percent, won the gold medal in the COVID-19 Olympics. Another way to score the death rate is to count COVID deaths per 100,000 people in a country, province, or state. By that measure, among the U.S. states, New York came in last with 55 per 100,000 people and California got the gold, with only 2.

These wild variations make one wonder whether the coronavirus has it in for certain groups. No—the numbers vary so much mostly because of how places count COVID deaths and how much testing has been done in different places. Some agencies report only people who tested positive before they died. If only people with severe symptoms get tested, naturally a higher portion of people who test positive will die and the virus will look super-deadly. That said, the virus does seem to have it in for older people. One reason why Italy's death rate was so high could be that it has the second-oldest population in the world.

How lethal is coronavirus? It sounds obvious but needs to be said: To know what a number means, you have to know what the counters included in their counts. The COVID death toll would be much higher if it included people who weren't infected but died for lack of medical care because the system was overwhelmed.

Numbers serve different purposes. New York governor Andrew Cuomo taught the nation how numbers can restore order and give us a sense of control when all seems lost. His daily press briefings looked like an online course in policy analysis, complete with slides of numbers, graphs, and always, the ominously rising curve of COVID cases. Step by step, Cuomo walked us through the numbers and showed how they can guide policy decisions. Unlike many in the Trump administration who worried that honest numbers would create fear and pandemonium, Cuomo believed that people would be reassured by his honesty and by knowing that he was facing the beast head-on.

In every briefing, he first reviewed the numbers, then showed how he used them to formulate a strategy.

"Let me take you through some facts today. . . . They are not encouraging," Cuomo said at the start of one briefing—but not before he thanked all the people who'd helped convert the Javits exhibition hall into an emergency hospital. "The increase in the number of cases continues unabated," he went on, with the curve on the screen behind him speeding upward past 5,700 cases. "We have the most sophisticated people you can get doing projections on this. . . . And what they're now seeing is that the rate of new infections is doubling every three days."

Numbers don't speak for themselves. To help us sense what such mind-boggling numbers mean, Cuomo used metaphors: "One of the forecasters said to me, 'We were looking at a freight train coming across the country. We're now looking at a bullet train.'" Talking about the curve, Cuomo frequently used the term "apex." Apex is math talk, so he likened the apex to a mountaintop. "We're all in search of the apex and the other side of the mountain, but we're still headed up the mountain." Often Cuomo compared the crisis to a war. In war, one must prepare to fight the enemy when and where it will be at its strongest. "The main battle is at the apex. The main battle is at the top of the mountain." Returning to the real world of crisis management, he explained that the apex is where "we have to be prepared to manage capacity," meaning, where we have to have enough hospital beds, staff, and equipment to handle all the cases.

In all the briefings Cuomo held before New York reached its apex, he displayed two sets of numbers—the projected numbers of beds, staff, and equipment that would be needed at the apex, and the amounts that hospitals currently had available. On March 25, 2020, forecasts predicted a need for 140,000 hospital beds and 53,000 ICU beds at the apex, but New York had only 53,000 hospital beds and 3,000 ICU beds. Without flinching, Cuomo told us in words what the numbers meant to him—and what they should mean to us: "These are troubling and astronomical numbers." The numbers were indeed frightening, but he laid out concrete plans for filling the gaps between how much was on hand now and how much would be needed at the apex.

Cuomo infused his numbers with credibility and clout. He used words and pictures along with the numbers to frighten people and motivate them to fight the enemy. He followed fear with reassurance and hope. He simplified the battle plan to an elementary arithmetic problem: Subtract a small number from a bigger number to find out the difference. He was open about where his numbers came from and the uncertainties they contained. He explained why, despite the uncertainty in the numbers, it's important to plan for the worst-case scenario. President Trump had questioned Cuomo's request for 30,000 ventilators. "I have a feeling that a lot of the numbers that are being said in some areas are just bigger than they're going to be," Trump said in a phone interview with Fox News's Sean Hannity. "I don't believe you need 40,000 or 30,000 ventilators." Cuomo gave a simple answer: "We don't need this stuff now. But we're planning for the battle at the apex."

Numbers told Cuomo *how much* equipment and personnel were needed, but they couldn't tell him *how to acquire* what was needed. For that, Cuomo used moral diplomacy. He always packaged the statistics in honesty, humility, and hope. He rarely took credit for improvements in the numbers. He asked people from out of state to donate equipment they weren't using and to send health care workers they could spare. He lost no opportunity to thank his staff and advisers, health and hospital workers, emergency personnel, the people who funded projections and donated equipment, and the volunteers and political leaders from around the country who came to New York's aid. Above all, in every press briefing he preached gratitude and reciprocity: "God bless America. Twenty-one thousand people have volunteered from out of state to come into New York State. I thank them. I thank their patriotism. I thank their dedication and passion to their mission of public health. . . . And New Yorkers will return the favor. . . . When your community needs help, New Yorkers will be there. And you have my personal word on that. . . . I will be the first one in my car to go wherever this nation needs help when we get past this. . . . I'll never forget how people across this country came to the aid of New Yorkers when they needed it."

President Trump taught the nation some other ways numbers can serve a leader's purposes. Months after experts warned that the coronavirus was going to be a pandemic, and two days before the World Health Organization officially declared it to be one, Trump likened it to a seasonal flu: "So last year 37,000 Americans died from the common Flu," he tweeted. "Nothing is shut down, life & the economy go on. At this moment, there are 546 confirmed cases of coronavirus, with 22 deaths. Think about that." Comparing the as-yet small coronavirus numbers with seasonal flu numbers was tantamount to saying coronavirus wasn't important enough to warrant strong government measures.

At times, Trump seemed to care mainly about how the numbers reflected on his leadership and how he could make sure they put him in a good light. When officials were deciding whether to let the *Grand Princess* cruise ship evacuate its infected and exposed passengers, Trump candidly admitted why he hoped the ship would be kept offshore: "I like the numbers being where they are. I don't need to have the numbers double because of one ship that wasn't our fault."

Later, when Trump was pushing to "re-open the country," he found himself trapped between two sets of numbers. He wanted the U.S. to score high on the number of people tested for the virus, but low on the number of people infected with it. Suddenly, in the midst of a meeting with Iowa's governor, he grasped his dilemma: "The media likes to say we have the most cases [of any country], but we do, by far, the most testing [or so he claimed]. If we did very little testing, we wouldn't have the most cases. So, in a way, by doing all of this testing, we make ourselves look bad." Cuomo was confident he could use bad numbers to strategize against the virus. Trump, by contrast, was unsure how to combine bad numbers with doing good.

Creating models to project numbers into the future helps us organize our knowledge—both the things we know and things we don't. National, state, and city leaders needed to anticipate how bad the pandemic would be, not only to plan for the apex, but also to decide how long to continue business shut-downs and social distancing. No one could know what the coronavirus was going to do, so leaders turned to mathematical models.

There's a saying in statistics: All models are wrong, but some are useful. All models are wrong because all models are based on assumptions. In the world of social science, assumption is a polite word for a guess. "Let's assume" is a way of saying, "Let's imagine that this, that, or the other is what's really going on in the world and then see what happens from there." Building a model is a bit like writing a play. A playwright imagines some characters with different personality traits, puts them in some imaginary circumstances, and then watches what happens. Playwrights often base their characters on people they've known and draw their circumstances from experiences they've had, or perhaps might like to have. Pandemic modelers take their assumptions from things they've observed in places where the virus has already done lots of damage.

Researchers around the world developed pandemic models to predict how the coronavirus would run its course. In the U.S., a model from the University of Washington became the most popular. Dubbed the IHME model, its projections were rosier than those of most other models. The original IHME model assumed that the growth of COVID cases in the U.S. would follow the growth pattern in China. But the U.S. isn't China. China took draconian measures to isolate people. The IHME model assumed that government stay-at-home orders would be as effective in the U.S. as they were in China. Not. The model assumed that all the American states would enact strict restrictions within a week after the model was released.

Three weeks later, seven states still had no statewide stay-at-home orders. The model assumed that the entire country would keep restrictions in place until summer. By late April, several states had begun to permit restaurants, schools, and nonessential businesses to reopen.

Dubious assumptions and all, models can be extremely useful. They enable officials to simulate different scenarios and anticipate challenges. The problem comes when a model's assumptions deviate far from reality. For that reason, smart leaders examine several models and choose ones that seem to best fit the situation. That approach, however, can lead to another problem. With multiple models at their disposal, officials can pick and choose among them to get numbers that suit their purposes. At one point in March when Trump was still reluctant to recommend social distancing, his advisers showed him a model from Imperial College in London that projected as many as 2.2 million deaths. The model got the desired response. Shortly after seeing it, Trump issued strong advice to limit gatherings to no more than 10 people.

A few weeks later, when the economic devastation had become painfully evident, the White House virus response team put out new, more optimistic projections, without explaining how it had arrived at the numbers. More optimistic projections would justify relaxing business closures and social distancing advice. Many observers feared that Trump's primary objective was to shore up his popularity by re-opening the economy and that his interest in stemming the pandemic was secondary. We can't know Trump's (or anyone's) motives for sure, but regardless of what they were, economic stability and public health are both legitimate goals. People choose models in part based on how much weight they give to different goals and how well a model serves their goals. Here's why even the best models are vulnerable to politics.

There's another reason why models are inaccurate, besides their built-in assumptions. Pandemic models work much like a Fitbit. A funny thing happens when you wear a Fitbit. It prods you to take more steps. The process of counting changes the count. Actually, the Fitbit Effect isn't so surprising, because people wouldn't wear a Fitbit if they weren't already motivated to exercise more. Pandemic models have similar feedback effects. They show how the curve would rise more slowly if leaders imposed stringent measures and residents cooperated. Seeing these happier numbers, some leaders issued new orders and more residents observed them. People changed their behavior to achieve better numbers, just as Fitbit wearers do.

It was just such positive feedback that made the numbers in New York and other places begin to slow down. Two weeks after the first IHME model, with new numbers in hand, the IHME modelers revised their death estimates downward. Instead of ranging from 100,000 to 240,000, the model predicted about 60,000 deaths. The new estimate was lower because it took into account the social distancing happening up to that point.

This is exactly how models are supposed to work. Modelers use new information to refine their assumptions. But models are built from imagination as well as facts, and so lend themselves to different interpretations. People of different political persuasions looked at their pandemic Fitbits and drew different conclusions from the new numbers. Dr. Anthony Fauci, the trusted voice of public health, said the numbers proved that social distancing was working—keep it up. Brian Kilmeade, a Fox News host, drew the opposite conclusion. He said the new numbers proved the models were drastically wrong. Shutting down the economy was an overreaction. It was time to "stand up this economy." "How many people are going to die as the economy goes flat on its back for three months?" he asked.

Trump had already raised this question in a screaming tweet: "WE CANNOT LET THE CURE BE WORSE THAN THE PROBLEM ITSELF!" Behind the bluster, Trump confronted the quintessential political conundrum: How should policymakers balance one kind of harm against another? How they should they balance damages caused by coronavirus against damages caused by shutting down the economy? How should any of us think about comparing so many different kinds of harms that affect different people in different ways?

These are questions of moral judgment. Numbers can help us think about moral questions, but they can't and shouldn't substitute for moral judgment. Nevertheless, Trump's tweet opened the door to answering the moral questions with numbers. Several economists came forward to offer, if not an answer, a method for calculating one: cost-benefit analysis. "Why is nobody putting some numbers on the economic costs of a month-long or a year-long shutdown against the lives saved?" asked one economic historian. Cost-benefit analysis sounds more scientific than it is. It hides difficult ethical choices in objective-seeming numbers. Like any form of counting, it frames issues by what analysts choose to count and what they leave out. And analysts can skew the answers by how they assign value to different things—and different people.

Some economists framed the corona problem as a trade-off between economic well-being and human health, as if the more we have of one, the less we can have of the other. That is a false choice. Several European governments imposed stringent business closures to slow the virus, but at the same time, they put up money to continue paying wages and salaries. In some places, governments underwrote businesses and guaranteed that workers would be able to return to their jobs when the pandemic was over. Where governments took those steps, economies didn't collapse so dramatically. Those government measures will certainly not be perfect, but they demonstrate an alternative to the cost-benefit framing. Governments don't have to ask, "Which costs less—protecting the economy or protecting human life?" They can do both.

Economists are used to thinking in dollar figures. For many of them, putting prices on human lives comes more easily than it does to most people. When they apply cost-benefit analysis to health and safety issues, they compare the costs of implementing protective regulations to the dollar value of lives that would be saved by the regulations. The field of cost-benefit analysis has a long tradition of devaluing older people. In the fall of 2019, the Council of Economic Advisers did a cost-benefit analysis of whether the federal government should invest in a vaccine development program to prepare for a possible pandemic. The study estimated that the economic costs of a pandemic would be about $3.8 billion, including the dollar value of lives lost.

Thus, an investment in vaccines up to $3.8 billion would be worth the cost. But here's the kicker: The council put a value of $12.3 million on each person aged 18 to 49, and a value of only $5.3 million on people over 65. The economic value of a vaccine development program (not to mention the social and emotional value) would have been much higher if the council hadn't deeply discounted older people's lives. There's nothing the least bit scientific or objective about pricing a grandmother at 43 percent of her grandchild. Precise numbers only serve to disguise the analysts' moral values.

Ageism reared its ugly head again as the media began publicizing how the virus affects different age groups. People over 70 are more vulnerable to contracting COVID-19 and more likely to die if they do get infected. Older people who have other health problems are even more likely to die from the virus. One 56-year-old lawyer tweeted what even the most die-hard cost-benefit proponents didn't dare say in public: "The fundamental problem is whether we are going to tank the entire economy to save 2.5% of the population which is (1) generally expensive to maintain, and (2) not productive." The lieutenant governor of Texas suggested that people over 70 should be willing to "take a chance on their survival" and let the economy to return to normal.

Governor Cuomo reacted to these suggestions with moral outrage: "My mother is not expendable. And your mother is not expendable. And our brothers and sisters are not expendable. And we're not going to accept a premise that human life is disposable. And we're not going to put a dollar figure on human life."

The corona pandemic poses a stark challenge to those who would hold firmly to equality as their moral standard. What are we supposed to do when there aren't enough life-saving resources to go around? How should a doctor or nurse decide which of several patients gets a ventilator? That is a terrible choice for anyone to make, especially health care workers imbued with the ethic of doing everything possible for each patient. Having to make these terrible choices leaves lasting psychic wounds, not only for medical personnel but also for everyone who must wonder, "How much is my mother worth? How much am I worth? Will people help us?"

Many cost-benefit analysts think of a life as being made up of *years of life* and would give priority to younger people in order to save more potential years. This formula isn't neutral. It privileges younger people over older people. Dan Wikler, a medical ethicist who has served on many task forces about rationing medical care, suggests a strategy of "saving the most lives possible," rather than coming up with decision rules about which people are more worth saving. If the goal is to save as many people as possible, priority should go to protecting and treating medical workers and other workers who are essential to maintaining civil order and providing food, water, and public utilities. Age should have no bearing. If there were only one ventilator, a 60- or 70-year-old health worker should get it rather than a 5- or 10-year-old child.

Putting the "save-the-most-lives" strategy into practice requires understanding how society functions and making thoughtful judgments. Doctors and nurses wouldn't be able to do much good without people who sterilize the equipment, wash the floors, push the gurneys around, empty the biohazard trash, and drive the buses and trains that take everyone to work. Under this strategy, medical workers might have to neglect some patients or even withdraw some patients from ventilators who have little chance of surviving. And it may well be that some younger people are more likely to survive than some older people, but a doctor wouldn't be taking ventilators from older people *because* they are worth less than younger people.

The save-the-most-lives strategy, though, still depends on arithmetic to resolve a moral dilemma, and that, to me, is troubling. The strategy is eerily reminiscent of a famous ethics problem. You must choose between letting a train run over 5 people who are tied to the tracks on one spur or throwing a switch to divert the train to another spur with only 1 person tied to the tracks. The save-the-most-lives rule is probably the right choice in that final desperate moment, but ethical leadership means taking action to prevent a situation from getting so desperate. Cuomo, along with many other governors and mayors, saw a chance to protect medical workers from having to make tragic choices by anticipating the peak need and stockpiling to meet it.

If we've reached the point of asking whose life is worth more, we have failed in our moral duties. Good moral reasoning means asking big questions, not posing puny arithmetic problems after you've let things get out of hand. What decisions and whose decisions put people in danger in the first place? Who has a chance to prevent tragic choices, and what can they do? In the train problem, I want to know which thugs tied the victims to the tracks and who could have stopped them. Preventing the crime is not only the more important ethical duty; it would save the most people, too.

The coronavirus pandemic inspired a lot of equality talk. We're all in this together. The virus makes no distinctions. Everyone's vulnerable to the disease and everyone's a potential transmitter. Those are nice sentiments, but they aren't quite true. By the end of March, a good two months into the U.S. pandemic, black, Hispanic, Native American, and immigrant communities knew the virus was hitting them hard, but there were few numbers to confirm their experience. "Why don't we know who the coronavirus victims are?" asked Ibram X. Kendi, author of *How to Be an Antiracist*. Of course, the victims and their families knew who they were, but "we"—the leaders, policymakers, scientists, and journalists tracking the pandemic—were in the dark.

We didn't know who the victims were because up to the time when Kendi asked the question, there were hardly any data. The CDC wasn't reporting race and ethnicity information. When a team of reporters examined racial disparities in COVID cases and deaths, it had to rely on the few state and local governments that had published victims' race. The figures confirmed what the trackers already knew or could have guessed. Blacks, Hispanics, and in some places Asians were infected and dying at rates much higher than their share of the population. In Louisiana, blacks made up 32 percent of the population but 70 percent of the COVID deaths. In Chicago, blacks made up 30 percent of the population but 68 percent of all COVID deaths. Similar disparities existed in Detroit, Milwaukee, and New

York. Skewed and sickening as the numbers were, they almost certainly undercounted disparities because even in the places that published race and ethnicity data, the data were missing from as many as half of the patient records. Besides, people in all these groups are less likely to have regular doctors and insurance, less likely to get tested, and so less likely to have their deaths counted as COVID deaths.

Kendi may not have been the first person to call for better victim data, but he laid out the political logic for seeking it. "Without racial data, we can't see whether there are disparities between the races in coronavirus testing, infection, and death rates. If we can't see racial disparities, then we can't see the racist policies behind any disparities and deaths. If we can't see racist policies, we can't eliminate racist policies."

As more data became available, they spotlighted more disparities that we already knew about if we cared to know. Native Americans are disproportionately victims. Poor people are disproportionately victims. Homeless people. Prisoners. Nursing home residents. People trapped in refugee camps and detention centers. Disproportions don't necessarily mean there's malice afoot. There's no reason to expect perfect correspondence between a group's share of the population and its share of trouble. However, vast disproportions in the distribution of something so devastating as COVID ought to make us wonder about the causes.

Pundits and politicians tended to describe the causes with euphemisms and stock tropes. The victims had "underlying conditions" or "pre-existing conditions." Their conditions were "rooted in longstanding economic and health inequalities." The victims were stuck in poverty and so couldn't afford healthy food and medical care. They tended to work in hard, dangerous jobs that took a toll on their health and didn't provide health insurance. While these explanations don't exactly blame the victim, they don't finger any guilty parties, either.

The worst-hit victim groups are like the people tied to the train tracks in the ethics dilemma, powerless to escape the virus bullet train. In this real-world ethics problem, though, we know a lot about how the victims got there. Whites had tied some of them to the tracks with outright discriminatory laws and policies in education, housing, jobs, health care, law enforcement, and financial credit. When courts struck down outright discriminatory laws, the same villains kept their victims tied to the tracks with under-the-radar discrimination. Enough legislators and lobbyists made sure that Congress wouldn't ever pass universal health insurance or sufficient safety nets to help the victims avoid those "underlying conditions." When states were given the opportunity to expand Medicaid under Obamacare, many of them—primarily former Jim Crow states—declined. Lack of health insurance isn't the only factor causing disparities in COVID-19 deaths, but state legislators who refused the Medicaid expansion refused an opportunity to help victims escape from the oncoming train.

When legislators, police, and prosecutors herd black and brown people into overcrowded, underfunded prisons, they lash their victims to the tracks. When state governments don't provide housing and support to people with developmental disabilities and mental illness, they lash them to the tracks. When employers don't provide paid sick leave, they lash their employees to the tracks. When they don't provide employees with personal protective equipment, they lash them to the tracks. When nursing home operators and health insurers starve the homes for funds, they lash residents to the tracks. When politicians threaten to deport immigrants and make them terrified to seek medical care, they lash them to the tracks.

Counting reveals some things and makes others invisible. Why don't we know who the coronavirus victims are? Because we count only what we care about. We don't count what we don't want to see. Coronavirus deaths are an acute manifestation of the social injustices we've come to accept as normal.

When the first racial disparity numbers for COVID deaths in Chicago came out, Mayor Lori Lightfoot gasped: "Those numbers take your breath away. They really do. This is a call to action."

Counting the disparities in COVID deaths is like doing an MRI on the body politic to reveal where injustice has already begun its deadly assault. But counting is just the beginning. Numbers are powerful tools only when we use them to write causal stories—to ask, "How did these numbers come about?" It's not enough to blame vagaries like "poverty" and "underlying conditions" as causes. It's not enough—in fact, it is heinously wrong—to "bring the economy back to normal," because normal ties the same victims to the tracks. It's not enough to identify victims of assault and come to their rescue just before the train hits them, or worse, leave them to die because we can't make up our collective mind what's the most cost-effective thing to do.

We have to stop the villains. And we have to ask ourselves what part we play in villainy.

The hero of Albert Camus' novel *The Plague* thinks his society is based on the death sentence. He sees the plague as a metaphor for how we are all complicit in the everyday death sentences our society carries out. "I learned," he says, "that I had had an indirect hand in the deaths of thousands of people; that I'd even brought about their deaths by approving of acts and principles which could only end that way. . . . Each of us has the plague within him; no one, no one on this earth is free from it. And I know, too, that we must keep endless watch on ourselves lest in a careless moment we breathe in somebody's face and fasten the infection on him."

Sometime during the early period of confusing precautionary advice, I asked my sister what she thought we should be doing about shopping, visiting friends, and wearing masks. She boiled it down to one simple rule: "Act is if you're already infected." That's a good rule for thinking about how we should behave as citizens. It's also a good rule for thinking about what we do when we count.

Many people think that when they count, they're gathering facts. In one of Cuomo's press briefings, he displayed some numbers about federal aid to states. Then he claimed, "I don't care what the news media tries to do to distort these facts. They are numbers, and they are facts, and they can't be distorted."

No, numbers are not facts. They don't exist in nature, silent and immobile until someone comes along and collects them. Numbers are created by people who count. When we decide what to count, we frame an issue as surely as a painter composes a scene. Our numbers embody the concerns, priorities, and values that guide us as we decide who or what belongs in the categories we're counting. These are moral choices. They can cause harm and they can do good. We should count as if we'll soon be infected by our own numbers. For in the end, what numbers do to others, they do to us as well.

Counting My Blessings

Numbers can't do justice to gratitude, so I'll have to count with words all the people who helped make this journey so fun and fulfilling.

First, there are the people who uttered one or two sentences that became my beacons.

My friend Dick Grant continues to tease me about words I wrote more than thirty years ago: "No number is innocent." That was the hunch that launched this book and over the years, Dick has fed my hunch with innumerable missives and conversations.

Three years ago, my colleagues at Aarhus University in Denmark listened as I explored where to take the hunch. Most everyone thought it should become a book, but Jørgen Elklit, who is, incidentally, an international expert on election integrity and counting ballots, smiled mischievously and asked, "So what?" His question challenged me from start to finish.

Jens Blom Hansen, also of Aarhus University, read the proposal for this book and said: "You've convinced me that numbers are weapons in political fights. But I guess that counting is here to stay, so how best do we live with it? Can you help us get counting right?" A tall order that both guided and haunted me.

Back home at Brandeis University, I presented the proposal to colleagues in the Politics Department. One person suggested that the answer to Jens's questions is "stick close to the raw data." While I fumbled for a polite reply, Jeremy Cynamon came to my rescue: "No! Isn't your argument that there's no such thing as a raw number?"

One day I tried out a passage on my cousin Penny MacCallum by reading aloud over the phone. After a minute or so, she stopped me in the midst of a gangly sentence: "I want to read your book. Don't make me have to work at it." I hope this book passes the Penny test.

Next, there are the people I counted on from Day One to read drafts, help me see the forest around the trees, and point out particularly interesting trees worth stopping to admire. Jens Blom Hansen and Jeremy Cynamon count in this group, too. So do Deb Chasman, Mike Doonan, Bob Kuttner, Lars Thorup Larsen, Shep Melnick, Laura Katz Olson, Marina Stanov, and Carolyn Tuohy.

Then come the people I met along the way who helped me make sense of their work and use it more accurately in telling my stories. In this group, I count Tony Affigne, Rebecca

Blouwolff, Alberto Cairo, Julia Dressel, Virginia Eubanks, Pamina Firchow, John Garcia, Samuel Gross, Kyle Jennison, Robert Norris, Spencer Piston, and Evelyn Whitehill.

My friends got into the spirit of the book so enthusiastically that our coffee dates, meals, and walks became colloquies on counting. For giving me a window into their numerating brains, I thank my brother David Stone, Judy Clementson, Deborah Garnick, Tore Johnson, Marty Krauss, Konnor Seymour and his grandmother Tami Geuser, and Sonia Wallenberg.

For being my all-around coaches and cheerleaders, tallyho to Marcia Angell, Cathryn Baird, Lindsay Evans, Charles Kaner, Lynn Mather, Mark Schlesinger, Carol Stack, and, once again, Marty Krauss and Carolyn Tuohy.

My family kept tabs on me even when I disappeared into the book for weeks at a time. Jeff, David, Judy, Lynne, Elizabeth, Ann Margaret, Jessica, and Samantha—thank you for being there while I was gone.

Last, there are the Ones—people in a category unto themselves.

Every book needs an in-house champion. Sasha Levitt, my editor at W. W. Norton, is that One. If there were a way to count enthusiasm for this book, hers would measure right up there with mine.

Bridget Miles is the One person in Muppet World who sees Cookie Monster and the Count as brilliant adult educators and made it possible for them to play a part in this book.

Vanessa Mulvey, my flute teacher, helps me feel Beat One inside the syncopated rhythms of music and life.

My friends, neighbors, and fellow citizens of Lempster, New Hampshire, show every day how One person can make a difference.

Several anonymous Ones belong in these acknowledgments, too, but I promised not to count them out loud.

There's One big problem with these acknowledgments. Each of my helpers should be counted in more than one category—which is exactly the problem that started me on this book.

Notes

Prologue: Of Two Minds

ix **Sir Charles Snow's phrase**: C. P. Snow, *The Two Cultures and the Scientific Revolution* (New York: Cambridge University Press, 1959).

x **Jacob Bronowski**: Jacob Bronowski, "The Creative Mind," in *Science and Human Values* (New York: Harper Torchbooks, 1965, rev. ed., originally published 1956), 1–24.

xii **"Stories aren't facts"**: Abby Patkin, "Town Mulls Pot Shop Restrictions," *Brookline Tab* (Brookline, MA), October 24, 2019.

xiii **"There is a world"**: Shawn Otto, *The War on Science: Who's Waging It, Why It Matters, What We Can Do About It* (Minneapolis, MN: Milkweed, 2016), 177.

xvi **"Stories are just data with a soul"**: Brené Brown, "The Power of Vulnerability," filmed June 2010 at TEDxHouston, TED video.

1. There's No Such Thing as a Raw Number

1 **Dr. Seuss**: *One Fish, Two Fish* (New York: Random House, 1960). I'm aware that Theodore Geisel, aka "Dr. Seuss," drew cartoons

with racist stereotypes during World War II. I think his later children's books celebrate diversity and acceptance of all people, as my reading of *One Fish, Two Fish* shows. And *One Fish, Two Fish* explains why diversity makes counting problematic better than any philosophy tomes.

5 **In the Sesame Street book**: *Sesame Street: 1 2 3 Count with Elmo* (Reader's Digest Children's Books, 2013).

7 **Until recently:** Kenneth Prewitt, *What Is* Your *Race?* (Princeton, NJ: Princeton University Press, 2013). The German doctor Johann Blumenbach named his categories with different labels than the ones the U.S. Census currently uses: Caucasian (white), Mongolian (Asian), Malay (Pacific Islander), Negro (African American or black), and American Indian (Native American).

7 **"A person of mixed white and Negro blood"**: Melissa Nobles, *Shades of Citizenship: Race and the Census in Modern Politics* (Stanford, CA: Stanford University Press, 2000), 72.

8 **"practically all Mexican laborers"**: Nobles, *Shades of Citizenship.* Emphasis and comment in brackets are mine.

8 **Gradually, the Census Bureau**: Margo Anderson, *The American Census*, 2nd ed. (New Haven, CT: Yale University Press, 2015); and Prewitt, *What Is* Your *Race*?

10 **case against Quinnipiac University**: R. Shep Melnick, *The Transformation of Title IX* (Brookings Institution Press, 2017).

11 "**Statistical method and statistical data are never ends in themselves**": Anderson, *The American Census*, 156.

11 **Darrell Huff**: Darrell Huff, *How to Lie with Statistics* (New York: W. W. Norton, 1954), 9.

13 **Rachel Kushner's short story**: Rachel Kushner, "Stanville," *New Yorker*, February 12 & 19, 2018, 77–85.

15 **Afghanistan's Independent Human Rights Commission**: Afghanistan Independent Human Rights Commission, *Annual Report 1394* (September 2016).

17 **Some experts think that if a country requires**: Matthew Coppedge et al., "Conceptualizing and Measuring Democracy: A New Approach," *Perspectives on Politics* 9, no. 2 (June 2011): 247–67.

18 **Pain is the most isolating experience there is**: Elaine Scarry, *The Body in Pain* (New York: Oxford University Press, 1985).

18 **Ronald Melzack**: Melzack wrote a retrospective account of how he came to develop his pain measure. Ronald Melzack, "The McGill Pain Questionnaire: From Description to Measurement," *Anesthesiology* 103, no. 1 (2005): 199–202. His original paper is Ronald Melzack and W. S. Torgerson, "On the Language of Pain," *Anesthesiology* 34 (1971): 50–59.

20 **"It's *The Presentation of Self in Everyday Life*"**: Erving Goffman, *The Presentation of Self in Everyday Life* (New York: Doubleday, 1959).

23 **Facebook founder Mark Zuckerberg is said to have lost**: Evan Osnos, "Ghost in the Machine," *New Yorker*, September 17, 2018, quoting a tweet by Nick Bilton, a technology writer at *Vanity Fair*.

23 **In some jobs, employees are "nickel and dimed"**: Barbara Ehrenreich, *Nickel and Dimed: On (Not) Getting By in America* (New York: Henry Holt, 2002).

24 **Here's how the U.S. government counts unemployment**: Janet L. Norwood and Judith M. Tanur, "Measuring Unemployment in the Nineties," *Public Opinion Quarterly* 58, no. 2 (Summer 1994): 277–94.

26 **Carroll Wright**: Alexander Keyssar, *Out of Work: The First Century of Unemployment in Massachusetts* (New York: Cambridge University Press, 1986), 1–3.

28 **"If you look for a job for six months"**: Glenn Kessler, "Donald Trump Still Does Not Understand the Unemployment Rate," *Washington Post*, December 12, 2016. The rally was on December 8, 2016.

29 **they test the wording of their questions on focus groups**: Norwood and Tanur, "Measuring Unemployment," 277–94. Janet Norwood served as director of the U.S. Bureau of Labor Statistics from 1979 to 1999 and oversaw a major reform of the survey questions. With guidance from cognitive scientists, she convened focus groups to test respondents' understanding of different question wordings.

29 **the government does tabulate the number of discouraged workers**: Steven E. Haugen, "Measures of Labor Underutilization from the Current Population Survey," BLS Working Paper #424, April 2009.

2. How a Number Comes to Be

33 **Piaget devised an experiment**: Piaget's experiment is reported by Jacques Mehler and Thomas G. Bever in "Cognitive Capacity in Very Young Children, *Science*, October 6, 1967, 141–42.

33 **MIT psychologists**: Mehler and Bever, "Cognitive Capacity," 141–42.

34 **Humans aren't the only creatures who count**: Stanislas Dehaene, *The Number Sense: How the Mind Creates Mathematics* (New York: Oxford University Press, 2011), chap. 1.

34 **Guppies given a choice**: Caleb Everett, *Numbers and the Making of Us: Counting and the Course of Human Cultures* (Cambridge, MA: Harvard University Press, 2017), 178–9.

34 **Pigeons and rats learn to count**: Damian Scarf, Harlene Hayne, and Michael Colombo, "Pigeons on Par with Primates in Numerical Competence," *Science*, December 23, 2011, 1554.

34 **A famous experiment**: Daniel J. Simons and Christopher F. Chabris, "Gorillas in Our Midst: Sustained Inattentional Blindness for Dynamic Events," *Perception* 28 (1999): 1059–74. The YouTube link is www.youtube.com/watch?v=IGQmdoK_ZfY, and there are many other versions of this experiment you can find as well. See also Arien Mack and Irvin Rock, *Inattentional Blindness* (Cambridge, MA: MIT Press, 1998).

35 **Health researchers have measured how breastfeeding**: Emily Oster, *Cribsheet: A Data-Driven Guide to Better, More Relaxed Parenting, from Birth to Preschool* (New York: Penguin, 2019), chap. 4.

35 **father's point of view**: Nathaniel Popper, "What Baby Formula Does for Fathers," *New York Times*, February 23, 2019.

36 **"cognitive biases"**: Daniel Kahneman and Amos Tversky, *Judgment under Uncertainty: Heuristics and Biases* (New York: Cambridge University Press, 1982) compiles many of their studies on cognitive biases. Tversky died in 1996. Kahneman describes their

joint work and later developments in cognitive biases in Kahneman, *Thinking Fast and Slow* (New York: Farrar, Straus and Giroux, 2011). It's delightfully readable and I draw most of my discussion from this book.

36 **percentage of various housework tasks**: Reported in Kahneman, *Thinking Fast and Slow*, 131.

37 **an "availability bias"**: Kahneman, *Thinking Fast and Slow*, 129–36.

37 **Which do you think kills more people—tornadoes or asthma?** Kahneman, *Thinking Fast and Slow*, 138. The importance of our emotions in judging frequency comes from research by Paul Slovic and his colleagues.

38 **Real estate agents**: Described in Kahneman, *Thinking Fast and Slow*, 124.

39 **genocide in Rwanda**: Kelly M. Greenhill, "Counting the Cost: The Politics of Numbers in Armed Conflict," in *Sex, Drugs, and Body Counts*, ed. Peter Andreas and Kelly M. Greenhill (Ithaca, NY: Cornell University Press, 2010), 127–58.

40 **In the run-up to 2008**: High finance is above my pay grade, but Michael Lewis gives a lucid, narrative explanation in *The Big Short: Inside the Doomsday Machine* (New York: W. W. Norton, 2010), especially in chap. 4. As Lewis explains, pay-for-grades wasn't the only reason bad bonds got good grades. Wall Street bond issuers figured out how to game the rating system so that their bonds appeared to be better than they were.

42 **President Clinton's big welfare reform**: There are many studies of how performance management stimulates creative counting. My favorites are Jason DeParle, *American Dream: Three Women, Ten Kids, and the Nation's Drive to End Welfare* (New York: Viking Penguin, 2004); and Joe Soss, Richard C. Fording, and Sanford F. Schram, *Disciplining the Poor* (Chicago: University of Chicago Press, 2011). Quotations in this paragraph are from the latter book.

42 **indicators to measure levels of violence**: I got the outlines of this story from Sally Engle Merry, *The Seductions of Quantification* (Chicago: University of Chicago Press, 2016), chap. 3. Merry sat in on committee meetings as an observer and did interviews over

several years. I have embellished her account with my own imaginings about what went on in the meeting rooms.

45 **The United Nations "Guidelines"**: United Nations, Department of Economic and Social Affairs, Statistics Division, *Guidelines for Producing Statistics on Violence Against Women—Statistical Surveys*, United Nations publication, 2014.

46 **Varieties of Democracy, or V-Dem for short**: V-Dem Institute, *Democracy for All? V-Dem Annual Democracy Report 2018* (Gothenburg, Sweden: University of Gothenburg, 2018). The country scores I cite are from the Liberal Democracy Index for 2018, on pp. 72–73. V-Dem runs separate surveys for 5 core aspects of democracy: elections, participation, deliberation, equality, and a catch-all "liberal democracy" that includes measures of election quality and civil liberties. The question on gender equality is from "Codebook: Exclusion and Social Media," Question 2.2 in the Exclusion module.

50 **"the degree to which they agree with other experts":** The method and quotations are in V-Dem Institute, *Democracy for All?*, 9.

51 **numbers at the top of the bureaucratic data chain**: The information on development statistics comes from Morten Jerven, *Poor Numbers: How We Are Misled by African Development Statistics and What to Do about It* (Ithaca, NY: Cornell University Press, 2013).

51 **Livestock numbers in Uganda**: Jerven, *Poor Numbers*, 87.

51 **Census Bureau has been the victim of partisan budget politics**: Margo Anderson, *The American Census: A Social History*, 2nd ed. (New Haven, CT: Yale University Press, 2015).

52 **"We fill it out on the way"**: From a confidential interview, October 19, 2018.

53 **"method of gap filling"**: Jerven, *Poor Numbers*, 22, based on the World Bank's document "Method of Gap Filling." I have paraphrased Jerven's quotation from the World Bank's manual: "a method based on the assumption that the growth of the variable [whatever is being measured] from a period for which the data exists has been the

same as the average growth for those other countries in the same region or income grouping, where data exists for both periods."

53 **course in Bird Counting 101**: "Bird Counting 101," eBird, February 23, 2012. Available at https://ebird.org/news/counting-101/. EBird is a citizen-science project based at the Cornell Lab of Ornithology, Cornell University.

57 **stir up the same kinds of hormones**: danah boyd, "Friends, Friendster and Top 8," *First Monday* 11, no. 12 (2006). boyd didn't study Facebook, but her ethnographic research with two other similar social network sites is equally applicable to Facebook.

58 **Recommender algorithms have figured out**: Zeynep Tufekci, "YouTube, the Great Radicalizer," *New York Times*, March 10, 2018; Paul Lewis, "'Fiction Is Outperforming Reality': How YouTube's Algorithm Distorts Truth," *The Guardian*, February 2, 2018.

58 **According to a former Google employee**: quoted in Lewis, "Fiction Is Outperforming Reality."

59 **a YouTube spokesperson**: quoted in Lewis, "Fiction Is Outperforming Reality."

59 **Police departments use algorithms to predict**: There are many articles about predictive policing. I base my analysis on the elegant mathematical demonstration of runaway feedback loops by Danielle Ensign et al., "Runaway Feedback Loops in Predictive Policing," *Proceedings of Machine Learning Research* 81 (2018): 160–71.

61 **Let's run the same kind of thought experiment**: Latanya Sweeney, director of the Data Privacy Lab in the Institute of Quantitative Social Science, Harvard University. Dr. Sweeney spun out this hypothetical in her keynote address at the Conference on Fairness, Accountability, and Transparency in Algorithms, at New York University, February 23, 2018.

3. How We Know What a Number Means

63 **Hans Rosling**: Hans Rosling was a Swedish public health professor who died in 2017. The story and quotations are from his posthumously published book, written with Ola Rosling and Anna Rosling Rönnlund, *Factfulness: Ten Reasons We're Wrong About the*

World—and Why Things Are Better Than You Think (New York: Flatiron Books, 2018), 19–20.

64 **Rosling wanted to learn how to swallow a sword**: Rosling, *Factfulness*, 1–2.

65 **Rebecca Blouwolff**: Rebecca Blouwolff, letter to the editor, *Brookline Tab* (Brookline, MA), April 4, 2019, and email correspondence with her in April 2019.

67 **According to the Census Bureau**: Quoted in Robert Pear, "Is Poverty a Condition or Is It a Definition?" *New York Times*, September 1, 1985.

67 **The U.S. method for measuring poverty**: I interviewed Mollie Orshansky in July 1993. Some of the material from my interview is published in Deborah Stone, "Making the Poor Count," *American Prospect* 17 (Spring 1994): 84–88. For historical material, I draw on Gordon Fisher, "The Development and History of the Poverty Thresholds," *Social Security Bulletin* 55 (Winter 1992): 3–14.

70 **The economy-plan budget didn't allow for**: Mollie Orshansky, "Counting the Poor," *Social Security Bulletin* 28, no. 1 (January 1965): 3–29.

71 **Fremstad offers some thought experiments**: Shawn Fremstad, "The Official U.S. Poverty Rate Is Based on Hopelessly Out-of-Date Metric," *Washington Post*, September 16, 2019.

71 **how much money one needs to stay afloat**: Richard Layard, *Happiness: Lessons from a New Science* (New York: Penguin, 2005), 4.

72 **as economist Robert Samuelson sees it**: Robert J. Samuelson, "Will the Real Poverty Rate Please Stand Up?" *Washington Post*, September 11, 2019.

72 **Samuelson thinks this way of counting income:** Samuelson, "Will the Real Poverty Rate Please Stand Up?"

74 **One student handbook advises**: Jane E. Miller, *The Chicago Guide to Writing about Numbers* (Chicago: University of Chicago Press, 2004), 5.

75 **One percent of families in the U.S. have more wealth**: Greg Leiserson, Will McGrew, and Rasha Koperam, "The Distribution of

Wealth in the United States and Implications for a Net Worth Tax," Washington Center for Equitable Growth, March 21, 2019.

75 **Blacks and Hispanics receive less screening**: Eric Peterson and Clyde W. Yancy, "Eliminating Racial Disparities in Cardiac Care," *New England Journal of Medicine* 360, no. 12 (March 19, 2009): 1172–74.

76 **researchers asked white people to read news articles**: Maureen A. Craig and Jennifer A. Richeson, "More Diverse Yet Less Tolerant? How the Increasingly Diverse Racial Landscape Affects White Americans' Racial Attitudes," *Personality and Social Psychology Bulletin* 40, no. 6 (2014): 1–12.

78 **upward-rising lines of good things:** Rosling's graphs of upward-rising lines for good things are in *Factfulness*, 62–63. For my graph "Research Publications per Year," I simplified Rosling's graph, "Science: Scholarly articles published per year," using rough numbers from his data.

78 **downward-falling lines of bad things**: Rosling's graphs of downward-falling lines for bad things are in *Factfulness*, 60–61. My graph of the decline in babies per woman simplifies Rosling's graph to show the overall downward trend.

80 **research on new drugs is riddled with conflicts of interest**: Marcia Angell, "Drug Companies and Doctors: A Story of Corruption," *New York Review of Books*, January 15, 2009.

80 **"It is simply no longer possible"**: Marcia Angell quoted in Carolyn Thomas, "NEJM Editor: 'No Longer Possible to Believe Much of Clinical Research Published," Ethical Nag, November 9, 2009.

80 **not the only story**: Paul Demeny, "Europe's Two Demographic Crises: The Visible and the Unrecognized," *Population and Development Review* 42, no. 1 (2016): 111–20; Anna Louie Sussman, "The End of Babies," *New York Times*, November 16, 2019.

81 **How big is Facebook?**: Evan Osnos, "Ghost in the Machine," *New Yorker*, September 17, 2018, 32–47.

81 **Climate change discussions are replete with numbers**: Bill McKibben, "How Extreme Weather Is Shrinking the Planet," *New Yorker*, November 26, 2018, 44–55.

82 **When Europeans first came to the New World**: W. Jeffrey Bolster, *The Mortal Sea* (Cambridge, MA: Belknap Press, 2012). The quotations are found on pp. 39–41, 43, 76, and 143.

83 **historian Jeffrey Bolster thinks**: Bolster, *The Mortal Sea*, 43.

83 **Bolster listens for the emotion between the lines**: Bolster, *The Mortal Sea*, 43.

84 **"Each family was allowed to embrace"**: The quotation is from Griselda San Martin, "The Door of Hope," an unpublished essay she sent me in an email on November 26, 2018. A report about her "Door of Hope" project can be seen at https://lens.blogs.nytimes.com/2016/11/14/a-song-of-love-and-longing-on-the-mexican-border/

85 **One of my favorite medical breakthroughs**: Raffi Khatchadourian, "Degrees of Freedom: A Scientist's Work Linking Minds and Machines Helps a Paralyzed Woman Escape Her Body," *New Yorker*, November 19, 2018.

85 **Gross domestic product**: My portrayal of the GDP relies mainly on Dirk Philipsen, *The Little Big Number: How GDP Came to Rule the World and What to Do about It* (Princeton, NJ: Princeton University Press, 2015).

88 **In the U.S., the GDP is announced with theatrics**: My description paraphrases Dirk Philipsen's in *The Little Big Number*, 147–48. According to Philipsen, other nations release their GDP figures with rituals that are similar, though perhaps not quite so melodramatic.

89 **"dark ages"**: Michael Boskin, as quoted in Philipsen, *The Little Big Number*, 149. Boskin made his remarks at a press conference at the U.S. Department of Commerce on December 7, 1999.

89 **"adrift in a sea"**: Paul Samuelson and William Nordhaus, *Economics*, 19th ed. (New York: McGraw-Hill Education, 2010), 390, as quoted in Philipsen, *The Little Big Number*, 148.

89 **"We tend to get what we measure"**: Gus Speth, "Toward a Post-Growth Society," *Yes Magazine*, July 2011, as quoted in Philipsen, *The Little Big Number*, 326.

90 **Genuine Progress Indicator**: Philipsen, *The Little Big Number*, 148–49, 224.

90 **Social Progress Index**: Scott Stern, Amy Wares, and Tamar Epner, *2018 Social Progress Index: Methodology Summary* (Social

Progress Imperative, 2018); *2018 Social Progress Index: Executive Summary* (Social Progress Imperative, 2018). For several of its component measures, the Social Progress Index uses scores from V-DEM, the Varieties of Democracy project we looked at in Chapter 2.

92 **Flint, Michigan, water crisis**: I base my account on Mona Hanna-Attisha, *What the Eyes Don't See: A Story of Crisis, Resistance, and Hope in an American City*, (New York: One World, 2018); and Anna Clark, *The Poisoned City: Flint's Water and the American Urban Tragedy* (New York: Metropolitan Books, 2018). My quotations are from these books unless noted.

93 **standards for dangerous levels of lead in blood**: Centers for Disease Control and Prevention, Childhood Lead Poisoning Prevention, "Blood Lead Levels in Children," page last updated by CDC on July 30, 2019, https://www.cdc.gov/nceh/lead/prevention/blood-lead-levels.htm (this document describes the name change from "level of concern" to "reference level"); and Centers for Disease Control and Prevention, "Recommended Actions Based on Blood Lead Level," page last updated by CDC on October 21, 2019, https://www.cdc.gov/nceh/lead/advisory/acclpp/actions-blls.htm (this document recommends only prevention, no medical treatment, for children with blood levels below 45).

96 **"The river water itself was not to blame"**: Quotation from Ashley Nickels, *Power, Participation, and Protest in Flint, Michigan: Unpacking the Policy Paradox of Municipal Takeovers* (Philadelphia: Temple University Press, 2019), 145. Nickels nicely shows how city leaders characterized the problem as a natural phenomenon, not the result of human decisions and actions.

97 **A group of Hurley physicians even voted**: Karen Bouffard, *Detroit News*, August 13, 2018. See also Hernán Gómez and Kim Dietrich, "The Children of Flint Were Not 'Poisoned,'" Opinion, *New York Times*, July 22, 2018.

4. How Numbers Get Their Clout

99 **front-page story**: Beth Teitell, "Boston's Clogged Arteries," *Boston Globe*, March 22, 2018.

101 **a local and customary affair**: Witold Kula, *Measures and Men* (Princeton, NJ: Princeton University Press, 1986).

102 **French scientists were cooking up the metric system**: Ken Alder, *The Measure of All Things* (New York: Free Press, 2002).

103 **"one weight, one measure"**: Quotations in this paragraph are from Alder, *The Measure of All Things*, 96, 252.

103 **"Gone are the countless daily opportunities"**: Kula, *Measures and Men*, 287.

104 **The U.S. Constitution begins by setting forth the structure of Congress**: U.S. Constitution, Article I, Section 2.

106 **slaves were counted as three-fifths of a person**: To be a little more precise, here's the language of Article 1, Section 2: a state's population "shall be determined by adding to the whole Number of free Persons, including those bound to Service for a Term of Years, and excluding Indians not taxed, three fifths of all other Persons." In plain English, indentured servants counted as part of a state's population. Indians who lived on reservations did not count, because they were exempt from taxes.

106 **These essays became known as *The Federalist Papers***: I use the Modern Library edition. *The Federalist: The Eighty-Five Essays Written by Alexander Hamilton, John Jay, James Madison* (New York: Random House, n.d.). There's some uncertainty as to whether Hamilton or Madison wrote Number 54, but most scholars credit Madison.

109 **Julia Dressel was a computer science and gender studies major**: Joseph Blumberg, "Julia Dressel '17 and Research That Went Viral," *Dartmouth News*, March 9, 2018.

109 **COMPAS, the algorithm**: COMPAS is an acronym for a suitably Orwellian name: "Correctional Offender Management Profiling for Alternative Services." The ProPublica study is by Julia Angwin et al., "Machine Bias: There's Software Used Across the Country to Predict Future Criminals. And It's Biased against Blacks," ProPublica, May 23, 2016.

110 **Julia Dressel wondered**: Dressel did the study as her senior honors thesis. She published the results in a coauthored article with her thesis advisor, Professor Hany Farid. Julia Dressel and Hany Farid, "The Accuracy, Fairness, and Limits of Predicting Recidivism," *Science Advances* 4, no. 1 (January 17, 2018): eaao5580.

112 **Public Safety Assessment**: This tool was developed by the Arnold Foundation, now called Arnold Ventures. Unlike COMPAS, the tool is available for free to jurisdictions that want to use it, and its data and formulas are publicly available. See "Public Safety Assessment: A Risk Tool That Promotes Safety, Equity, and Justice," Arnold Ventures, August 2017.

113 **tend to go by their intuitions and the rules of thumb**: The quotations about judges in this paragraph are from a TED talk by Anne Milgram, "Why Smart Statistics Are the Key to Fighting Crime," filmed October 2013 in San Francisco, CA, TED video. Milgram helped develop the Public Assessment Safety tool for the Arnold Foundation.

115 **"That's why I love predictive analytics"**: Rachel Berger, director of a child-abuse research center at Children's Hospital, Pittsburgh, quoted in Dan Hurley, "Can an Algorithm Tell When Kids Are in Danger?" *New York Times Magazine*, January 2, 2018.

115 **Department of Human Services in Allegheny County**: What I call "the child-welfare department" is formally called the Department of Children, Youth and Families and is part of the Department of Human Services. The algorithm is formally called the Allegheny Family Screening Tool.

116 **Every time a call comes in to the department**: My description of the screening process is drawn from Alexandra Chouldechova, Emily Putnam Hornstein, Diana Benavides-Prado, Olesksandr Fialko, and Rhema Vaithianathan, "A Case Study of Algorithm-Assisted Decision Making in Child Maltreatment Hotline Screening Decisions," *Proceedings of Machine Learning Research* 81 (2018): 1–15; and Virginia Eubanks, *Automating Inequality: How High-Tech Tools Profile, Police, and Punish the Poor* (Berkeley: University of California Press, 2017), chap. 4. Evelyn Whitehill and Kyle Jennison, both in the Allegheny County Department of Human Services, generously read drafts and answered questions to help me understand how the algorithm works and its place in the agency's decision making.

116 **Virginia Eubanks**: Eubanks, *Automating Inequality*, chap. 4.

118 **The category of "mandatory investigation"**: Allegheny County Department of Human Services, "Allegheny Family Screening Tool: Frequently Asked Questions," updated August 2018. Recently, the department has changed the language and rules for required investigations. It no longer uses the word "mandatory," but some cases still need a supervisor's permission to override a required investigation. Email correspondence from Evelyn Whitehill, Allegheny County Department of Human Services, Office of Analytics, Technology and Planning, November 18, 2019.

118 **Screeners tend to rely on their own scores**: Chouldechova et al., "A Case Study."

118 **evaluation commissioned by the department**: *Allegheny County Predictive Risk Modeling Tool Implementation: Process Evaluation*, Hornby Zeller Associates, January 2018. Even though the consultant's samples included only 16 or 18 screeners, the report gives percentages rather than the actual numbers of screeners who answered one way or another. Since percentages can be misleading when based on such a small sample, I calculated the actual numbers from the percentages.

119 **"typically there's something you're missing"**: Quotation from Eubanks, *Automating Inequality*, 141.

119 **"I call it someone's opinion"**: Erin Dalton quoted in Hurley, "Can an Algorithm Tell."

119 **Someone on Dalton's staff**: Evelyn Whitehill explained what she thought Dalton meant in her statement, in email correspondence on November 18, 2019.

119 **"We definitely oversample the poor"**: Dalton quoted in Eubanks, *Automating Inequality*, 158.

120 **"All of the data . . . is biased"**: Erin Dalton quoted in Hurley, "Can an Algorithm Tell."

121 **improved the department's accuracy**: Jeremy D. Goldhaber-Fiebert and Lea Prince, "Impact Evaluation of a Predictive Risk Modeling Tool for Allegheny County's Child Welfare Office," Allegheny County Analytics, March 20, 2019.

121 **Prediction #1 comes true**: According to Kyle Jennison, lead analyst in the Department of Human Services, as of mid-2019, the

algorithm no longer uses future hotline calls as one of its predictors. To its credit, the analytics team constantly evaluates and modifies the tool. Apparently, risk of foster care placement was a better predictor of abuse than referral calls.

121 **asked two ethicists to conduct an ethical review:** Tim Dare and Eileen Gambrill, "Ethical Analysis: Predictive Risk Models at Call Screening for Allegheny County," April 2017. This report is available only online at: https://www.alleghenycountyanalytics.us/index.php/2019/05/01/developing-predictive-risk-models-support-child-maltreatment-hotline-screening-decisions/ Tim Dare, one of the authors commissioned by the Allegheny County Department of Human Services, had already given an ethical green light to New Zealand's Ministry of Social Development when it was deciding whether to adopt a child maltreatment algorithm developed by the same people who designed the Allegheny tool. See Tim Dare, *Predictive Risk Modelling and Child Maltreatment: An Ethical Review* (Auckland, New Zealand: University of Auckland, September 25, 2013).

122 **"While it is true that predictive risk modeling tools will make errors":** Dare and Gambrill, "Ethical Analysis," p. 4.

122 **"ethically appropriate":** Dare and Gambrill, "Ethical Analysis," p. 9.

122 **the algorithm can "mis-categorize":** Dare and Gambrill, "Ethical Analysis," p. 4.

122 **"individuals identified as at high risk must not be treated":** Dare and Gambrill, "Ethical Analysis," p. 5.

123 **"The fact that any intervention is designed to assist":** Dare and Gambrill, "Ethical Analysis," p. 6. The complete sentence reads: "The fact that any intervention is designed to assist gives grounds to think the model is not vulnerable to the legitimate concerns generated by the existence of disparities in data used in punitive contexts."

126 **Pay-for-performance has ruled the roost**: Diane Ravitch, *The Death and Life of the Great American School System: How Testing and Choice Are Undermining Education* (New York: Basic Books, rev. ed. 2016). Ravitch was one of the supporters and architects of

the system but came to feel she had created a monster. Her book is both a trove of information about American education reform and a Frankenstein-esque horror story.

126 **typical student takes 112 standardized tests**: Lyndsey Layton, "Study Says Standardized Testing Is Overwhelming Nation's Public Schools," *Washington Post*, October 24, 2015.

127 **a shrine to numbers**: My understanding of value-added modeling is informed by Kevin Carey, "The Man Who Measured Teachers," *New York Times*, May 21, 2017; Ravitch, *The Death and Life*; and the many news articles cited in this chapter.

128 **Sheri Lederman**: Valerie Strauss, "Controversial Teacher Evaluation Method Is on Trial—Literally—and the Judge Is Not Amused," *Washington Post*, August 15, 2015; and Valerie Strauss, "Judge Calls Evaluation of N.Y. Teacher 'Arbitrary' and 'Capricious' in Case against New U.S. Secretary of Education," *Washington Post*, May 10, 2016.

128 **even the custodians**: Valerie Strauss, "D.C. Custodial Staff Were Evaluated by Student Test Scores. Really," *Washington Post*, April 16, 2013.

128 **half of a biology teacher's assessment**: Valerie Strauss, "Lawsuit: Stop Evaluating Teachers on Test Scores of Students They Never Taught," *Washington Post*, April 15, 2015.

128 **art teachers' scores**: Valerie Strauss, "How Is This Fair? Art Teacher Is Evaluated by Students' Math Standardized Test Scores," *Washington Post*, March 25, 2015.

129 **Not long into the pay-for-performance regime**: Examples in this paragraph and the next are from Ravitch, *The Death and Life*; and Jerry Z. Muller, *The Tyranny of Metrics*, (Princeton, NJ: Princeton University Press, 2018), chap. 8.

130 **lowering the thresholds for passing**: Ravitch, *The Death and Life*, 113–14.

130 **teachers gave diplomas to students who didn't meet the graduation requirements**: Emma Brown, Valerie Strauss, and Perri Stein, "It Was Hailed as the National Model for School Reform. Then the Scandals Hit." *Washington Post*, March 10, 2018.

132 **Jerry Muller traces pay-for-performance reforms:** Muller, *Tyranny of Metrics*.

137 **"Smart statistics are the key to fighting crime"**: From the title of a TED talk by Anne Milgram, "Why Smart Statistics Are the Key to Fighting Crime."

137 **"Value-added analysis offers the closest thing"**: Jason Felch, Jason Song, and Doug Smith, "Who's Teaching LA's Kids?" *Los Angeles Times*, August 14, 2010.

137 "**Value-added also may serve"**: Ted Hershberg, Virginia Adams Simon, and Barbara Lea-Kruger, "The Revelations of Value-Added: An Assessment Model That Measures Student Growth in Ways That NCLB Fails to Do," *School Administrator* 61, no. 11 (December 2004).

138 **"mathematical intimidation"**: John Ewing, "Mathematical Intimidation: Driven by the Data," *Notices of the American Mathematical Society* 58, no. 5 (May 2011): 667–73.

138 **Think about the way arithmetic is taught**: Paul Lockhart, *A Mathematician's Lament: How School Cheats Us Out of Our Most Fascinating and Imaginative Art Form* (New York: Bellevue Literary Press, 2009).

5. How Counting Changes Hearts and Minds

141 **Erling Kagge loves to walk**: Erling Kagge, *Walking: One Step at a Time* (New York: Pantheon Books, 2019), 139–42.

142 **nutritionism**: Michael Pollan, "Unhappy Meals," *New York Times*, January 28, 2007; and Michael Pollan, *In Defense of Food* (New York: Penguin, 2008). The quotation about food labeling is from "Unhappy Meals."

143 **"Food is also about pleasure"**: Pollan, *In Defense of Food*, 8.

144 **Some research suggests:** Jordan Etkin, "The Hidden Cost of Personal Quantification," *Journal of Consumer Research* 42, no. 6 (April 2016): 967–84; Alfie Kohn, *Punished by Rewards: The Trouble with Gold Stars, Incentive Plans, A's, Praise and Other Bribes* (New York: Houghton Mifflin, 1993). Etkin, whose research about "counting more and enjoying it less" is widely cited, acknowledges that if you're engaging in an activity to meet your own goal, then measurement

can "have some benefits for enjoyment." Quoted in Robinson Meyer, "The Quantified Welp," Atlantic.com, February 28, 2016.

144 **"outsourcing self-management"**: Frank Pasquale, "How Fitness Trackers Make Leisure More Like Work," Atlantic.com, March 2, 2018.

144 **"unwind more efficiently"**: Quotations from Jennifer Breheny Wallace, "For a Relaxing Vacation, Look to the Data," *Wall Street Journal* (online), August 26, 2016, quoting Steve Jonas.

145 **"the pulse of democracy"**: George Gallup and Saul Forbes Rae, *The Pulse of Democracy: The Public-Opinion Poll and How It Works* (New York: Simon and Schuster, 1940).

145 **"high-tech equivalent"**: Quotation from Robert S. Erikson and Kent L. Tedin, *American Public Opinion* (New York: Routledge, 2016), 5.

145 **National Opinion Research Center**: Quotation from the organization's website, www.norc.org/Research/Topics/Pages/society,-media,-and-public-affairs.aspx.

147 **The interviewer asks many questions**: The questions in this section are drawn from the American National Election Study, various years. I found the questions by using the site's "Continuity Tool" and searching under Part VIII Social Groups, section on "Group Stereotypes": http://isr-anesweb.isr.umich.edu/ANES_Data_Tools/ContinuityTool_v2.html?v=04112019 - viii-social-groups.

148 **1. Peaceful . . . 7. Violent:** This question about violence and social groups was asked in 1992 and 2016. In 2016 the same question was asked about Christians and Muslims. A related question reads: "How well does the word *violent* describe [Muslims?] [Mormons]?" This question was asked only in 2012.

148 **1. Intelligent . . . 7. Unintelligent:** This question on intelligence and social groups was asked in 1992, 1996, 2000, 2004, 2008, and 2012; it asked about whites, blacks, and Hispanics, and asked about Asian Americans in all these years except 1996.

149 **derogatory images of blacks**: James A. Morone, *Hellfire Nation* (New York: Basic Books, 2003); Ibram X. Kendi, *Stamped from the Beginning* (New York: Nation Books, 2015).

152 **Survey questions can activate ideas that are already there**: Thanks to Spencer Piston, political scientist at Boston University,

for teaching me about the public opinion field and pointing out the difference between putting new ideas in people's heads and making available what's already there.

155 **Let's try some more questions**: The questions in this section are from the American National Election Study 2016 unless otherwise noted.

155 **"What do you think are the chances"**: This question is from an earlier American National Election Study, reported in Donald R. Kinder and Lynn M. Sanders, *Divided by Color: Racial Politics and Democratic Ideas* (Chicago: University of Chicago Press, 1996), 54. For comparison's sake, in the 2016 survey this question is worded: "How likely is it that many whites are unable to find a job because employers are hiring minorities instead?" (Possible answers are "extremely likely," "very likely," "moderately likely," "slightly likely," "not at all likely.") Note that the current question is worded less personally ("many whites" instead of "you or someone in your family") and less pointedly (without the phrase "unqualified black," and substituting "minorities" for "blacks").

155 **"Do you think immigrants"**: This question is from a Pew Research Center survey conducted in March 2016. Bradley Jones, "Americans' Views of Immigrants Marked by Widening Partisan, Generational Divides," Pew Research Center, April 15, 2016. The Pew Research Center has been asking this question for over 20 years.

159 **Most people . . . don't know the half of what government does for them**: Suzanne Mettler, *The Submerged State* (Chicago: University of Chicago Press, 2011).

160 **techniques to reduce unconscious bias**: Much of this work is done by Project Implicit, a nonprofit founded by the University of Virginia psychologist Brian Nosek. The experiments I describe are reported in Calvin K. Lai and 21 coauthors, including Nosek, "Reducing Implicit Racial Preferences: I. A Comparative Investigation of 17 Interventions," *Journal of Experimental Psychology: General* 143, no. 4 (2014): 1765–85.

164 **taking advice from some Republican strategists**: Michael Wines, "Deceased G.O.P. Strategist's Hard Drives Reveal New

Details on the Census Citizenship Question," *New York Times*, May 30, 2019.

164 **feared the Census Bureau would send their information directly to ICE**: Census Bureau, Center for Survey Measurement, "Memorandum for Associate Directorate for Research and Methodology," September 20, 2017.

165 **Census Bureau experts weighed in against it**: Michael Wines, "Census Bureau's Own Expert Panel Rebukes Decision to Add Citizenship Question," *New York Times*, March 30, 2018.

165 **"Your responses to the 2020 Census are safe"**: U.S. Census Bureau, *The 2020 Census and Confidentiality*, Document D-1254 (Washington, DC: U.S. Census Bureau, n.d.). Boldface in original.

166 **played an important part in creating a Hispanic identity**: Two people contributed mightily to my education on the census and Hispanic identity: Tony Affigne, political science professor at Providence College, and John Garcia, Research Professor Emeritus, University of Michigan. Affigne reminded me that the Census Bureau doesn't completely control its own questions, as we saw when Trump's Commerce secretary ordered the bureau to add a citizenship question in 2020. In addition to several books and articles about the census, I relied heavily on two books about Hispanic identity: G. Christina Mora, *Making Hispanics: How Activists, Bureaucrats and Media Constructed a New American* (Chicago: University of Chicago Press, 2014); and Clara E. Rodriguez, *Changing Race: Latinos, the Census, and the History of Ethnicity in the United States* (New York: New York University Press, 2000).

168 **President Hoover deported**: Mora, *Making Hispanics*, 110.

168 **He invited Hispanic civic group leaders**: Mora, *Making Hispanics*, 94–95, 110–14.

169 **The Hispanic or Latino identity caught on**: Luis Ricardo Fraga et al., *Latino Lives in America* (Philadelphia: Temple University Press, 2010), chapter 7. In 2002, Cuban Americans were the only Spanish-origin group with fewer than 80 percent saying they ever call themselves Hispanic or Latino; among them the number was 73 percent.

170 **"Some other race" is now the third-largest**: Sowmiya Ashok, "The Rise of the American 'Others,'" Atlantic.com, August 27, 2016.

171 **why they answer census questions as they do**: Clara E. Rodriguez, "Race, Culture, and Latino 'Otherness' in the 1980 Census," *Social Science Quarterly* 73, no. 4 (1992): 930–37; and Rodriguez, *Changing Race*.

172 **a study of 31 countries**: Evan S. Lieberman and Prerna Singh, "Census Enumeration and Group Conflict: A Global Analysis of the Consequences of Counting," *World Politics* 69, no. 1 (2017): 1–53.

173 **Innocence work has become a veritable social movement**: My history of the innocence movement draws mostly from Robert Norris, *Exonerated: A History of the Innocence Movement* (New York: New York University Press, 2017). Quotations are from this book unless otherwise noted.

174 **We can't know why their estimates rose**: Samuel Gross, "How Many False Convictions Are There? How Many Exonerations Are There?" in C. Ronald Huff and Martin Killias eds., *Wrongful Convictions and Miscarriages of Justice* (New York: Routledge, 2013).

174 **conference in Chicago**: My description of the conference and the quotations arc from Norris, *Exonerated*, 73–75.

175 **more crimes would be committed by the real perpetrator**: Subsequent research bears out this claim. Almost all of the true perpetrators identified by DNA evidence had committed more crimes while their innocent counterpart was in prison, crimes that included homicides and sex offenses. Robert Norris et al., "The Criminal Costs of Wrongful Convictions," *Criminology and Public Policy*, forthcoming.

177 **"milestone reports" to publicize their political successes**: "Indiana Becomes 34th State to Pay Exonerees for Wrongful Incarceration," Innocence Project, May 1, 2019; "Oklahoma Becomes 25th State to Require Recording of Interrogations," Innocence Project, May 14, 2019. There is a wealth of information on other legal reforms available on the organization's website, www.innocenceproject.org.

177 **"We now often say we do two things at the Registry"**: The quotations are from phone conversations and email correspondence with Samuel Gross in June and December 2019.

178 **no such person as "the innocent man convicted"**: Judge Learned Hand, United States District Court, Southern District

of New York, *United States v. Garsson* (1923), quoted in Gross, "How Many False Convictions Are There?" Judge Hand's actual words as quoted by Gross were: "Our [criminal] procedure has always been haunted by the ghost of the innocent man. It is an unreal dream."

6. The Ethics of Counting

179 **One of the most important American civil rights cases**: *Masterpiece Cakeshop, Ltd. v. Colorado Civil Rights Commission*, 584 U.S. ___(2018). My description of the legal significance of the case is based on David Cole, "Let Them Buy Cake," *New York Review of Books*, December 7, 2017.

180 **"I'm being forced to use my creativity":** The baker, Jack Phillips, was quoted in Adam Liptak, "Cake Is His 'Art.' So Can He Deny One to a Gay Couple?" *New York Times*, September 16, 2017.

185 **how point systems work**: My sketch draws together some factors used by different universities. Most large schools use commercial software packages such as Academic Analytics (https://academicanalytics.com). See Colleen Flaherty, "Rutgers Graduate School Faculty Takes a Stand against Academic Analytics," Inside Higher Ed, May 11, 2016.

188 **If Konnor dreams of playing baseball**: Malcolm Gladwell, *Outliers: The Story of Success* (New York: Little, Brown, 2008), chap. 1.

189 **The company that runs the SATs recently decided**: Anemona Hartocollis, "SAT's New 'Adversity Score' Will Take Students' Hardships into Account," *New York Times*, May 16, 2019; Thomas Chatterton Williams, "The SAT's Bogus 'Adversity Score,'" *New York Times*, May 17, 2019; Richard D. Kahlenberg, "An Imperfect SAT Adversity Score Is Better Than Just Ignoring Adversity," Atlantic .com, May 25, 2019.

192 **If Konnor gets a job driving for Uber**: My portrait of Uber is based on Alex Rosenblat, *Uberland: How Algorithms Are Rewriting the Rules of Work* (Berkeley: University of California Press, 2018).

194 **Automated essay scoring:** Tovia Smith, "More States Opting to 'Robo-Grade' Student Essays By Computer," NPR, June 30, 2018;

and Peter Greene, "Automated Essay Scoring Remains an Empty Dream," *Forbes*, July 2, 2018.

195 **"There's opportunity to fix it"**: Virginia Eubanks, *Automating Inequality* (Princeton, NJ: Princeton University Press, 2017), 167.

195 **"People who have gone through a trauma"**: Eubanks, *Automating Inequality*, 62.

201 **Everyday Peace Indicators**: Pamina Firchow, *Reclaiming Everyday Peace: Local Voices in Measurement and Evaluation after War* (New York: Cambridge University Press, 2018). I drew the examples for Afghanistan from the Everyday Peace Indicators Codebook for Afghanistan, available at https://everydaypeaceindicators.org/codebooks/.

202 **one of the most famous puzzles about ethics**: For all versions of the trolley problem, I draw from David Edmonds, *Would You Kill the Fat Man: The Trolley Problem and What Your Answer Tells Us about Right and Wrong* (Princeton, NJ: Princeton University Press, 2014).

209 **whenever government agencies propose a new regulation**: Clinton issued Executive Order 12866 in 1993. In 2011, Obama issued Executive Order 13563. My discussion of cost-benefit analysis relies heavily on two sources, the first strongly supportive and the second strongly critical. Cass R. Sunstein, *Valuing Life: Humanizing the Regulatory State* (Chicago: University of Chicago, 2014); and Frank Ackerman and Lisa Heinzerling, *Priceless: On Knowing the Price of Everything and the Value of Nothing* (New York: New Press, 2004).

211 **Cost-benefit analysis is an exercise in imagination**: Cost-benefit analysts recognize this problem, so they've come up with some counting rules. For example, we'll consider only ripples that happen during the first 5 years or only the benefits and harms to people, not to plants, animals, and soil. Rules like these make the problem manageable, but they're arbitrary.

212 **the more people who die early**: W. Kip Viscusi, "Smoke and Mirrors: Understanding the New Scheme for Cigarette Regulation," *Brookings Review* (Winter 1998): 14–19; and W. Kip Viscusi,

"Cigarette Taxation and the Social Consequences of Smoking," *Tax Policy and the Economy* 9 (1995): 51–101.

213 **Prison Rape Elimination Act**: I draw this example and the numbers from Sunstein, *Valuing Life*, 75. Another helpful take on "willingness-to-pay" is Lisa Heinzerling, "Cost-Benefit Jumps the Shark," Georgetown Law Faculty Blog, June 13, 2012.

217 **dedicated the cemetery at Gettysburg**: Abraham Lincoln, "The Gettysburg Address" (Bliss copy), November 19, 1863.

217 **Toni Morrison called on Lincoln**: Toni Morrison, Nobel Lecture, delivered December 7, 1993.

Epilogue: Counting Goes Viral

220 **identified some likely suspects**: Carolyn Y. Johnson and Laurie McGinley, "What Went Wrong with the Coronavirus Tests in the U.S.," *Washington Post*, March 7, 2020; Olga Khazan, "The 4 Key Reasons the U.S. Is So Behind on Coronavirus Testing," *Atlantic .com*, March 13, 2020; Rebecca Ballhouse and Stephanie Armour, "Health Chief's Early Missteps Set Back Coronavirus Response," *Wall Street Journal*, April 22, 2020.

221 **How far can a cough project**: Knvul Sheikh, James Gorman, and Kenneth Chang, "Stay 6 Feet Apart, We're Told. But How Far Can Air Carry Coronavirus?" *New York Times*, April 14, 2020.

221 **newly jobless**: "In 4 Weeks, 22 Million Americans Have Lost Their Jobs," *New York Times*, April 16, 2020.

221 **FEMA personnel**: "FEMA, Racing to Provide Virus Relief, Is Running Short on Front-line Staff," *New York Times*, April 4, 2020.

221 **bottles of hand sanitizer**: Jake Nicas, "He Has 17,700 Bottles of Hand Sanitizer and Nowhere to Sell Them," *New York Times*, March 14, 2020.

221 **condoms**: Sam Reeves, "Virus May Spark 'Devastating' Global Condom Shortage," *Barron's*, April 8, 2020.

221 **cyberscams**: Joseph Marks, "The Cybersecurity 202: Coronavirus Pandemic Unleashes Unprecedented Number of Cyberscams," *Washington Post*, April 8, 2020.

222 **Initially the CDC counted only deaths**: Emma Brown, Beth Reinhard, and Reis Thebault, "Which Deaths Count Toward the

COVID-19 Death Toll? It Depends on the State," *Washington Post*, April 16, 2020.

222 **The day after a reporter broke the New York City story**: Gwynne Hogan, "Staggering Surge of NYers Dying in Their Homes Suggest City Is Undercounting Corona Virus Fatalities," *Gothamist*, April 7, 2020; Gwynne Hogan, "Death Count Expected to Soar as NYC Says It Will Begin Reporting Probable COVID Deaths in Addition to Confirmed Ones," *Gothamist*, April 9, 2020; J. David Goodman and William K. Rashbaum, "N.Y.C. Death Toll Soars Past 10,000 in Revised Virus Count," *New York Times*, April 14, 2020.

223 **some agencies now count people who probably died**: Brown, Reinhard, and Thebault, "Which Deaths Count?"; Martha Henriques, "In Italy, the Death Rate from COVID-19 Is More Than 10 Times Greater Than in Germany" BBC.com, April 1, 2020.

223 **The most common way**: The method of calculating fatality rate and the figures are based on Henriques, "In Italy, the Death Rate," and Amy Harmon, "Why We Don't Know the True Death Rate for COVID-19," *New York Times*, April 17, 2020. Epidemiologists have other ways to measure the death rate but I'm keeping it simple here.

224 **One reason why Italy's death rate was so high**: Harmon, "Why We Don't Know the True Death Rate."

225 **"Let me take you through some facts today"**: Governor Andrew Cuomo's press briefing, March 25, 2020.

225 **"One of the forecasters said to me"**: Cuomo's press briefing, March 25, 2020

225 **"We're all in search of the apex"**: Cuomo's press briefing, March 31, 2020.

225 **"The main battle is at the apex"**: Cuomo's press briefing, March 31, 2020.

225 **the apex is where "we have to be prepared"**: Cuomo press briefing, March 25, 2020.

226 **"I have a feeling that a lot of the numbers"**: Trump, quoted in William Wan and Aaron Blake, "Coronavirus Modelers Factor in a New Public Health Risk: Accusations Their Work Is a Hoax," *Washington Post*, March 27, 2020.

226 **"We don't need this stuff now"**: Cuomo's press briefing, March 31, 2020.

227 **"God bless America"**: Cuomo's press briefing, April 2, 2020.

227 **"So last year 37,000 Americans died"**: Trump tweet on March 9, 2020, quoted in Ashley Parker, Josh Dawsey, and Yasmeen Abutaleb, "For Trump, the Coronavirus Is All about the Numbers—and They Don't Look Good," *Washington Post*, March 12, 2020.

228 **At times, Trump seemed to care**: Parker, Dawsey, and Abutaleb, "For Trump, the Coronavirus Is All about the Numbers."

228 **"I like the numbers being where they are"**: Trump made this statement during his visit to the CDC on March 6, 2020; quoted in Parker, Dawsey, and Albutaleb, "For Trump, the Coronavirus Is All about the Numbers."

228 **"The media likes to say"**: Philip Bump, "The Sharp Hypocrisy of the White House Position on Testing," *Washington Post*, May 6, 2020.

229 **a model from the University of Washington**: The model was designed by the Institute for Health Metrics and Evaluation at the University of Washington, hence the acronym IHME.

229 **The original IHME model**: William Wan and Aaron Blake, "Coronavirus Modelers Factor in a New Public Health Risk"; William Wan, Josh Dawsey, Ashley Parker, and Joel Achenbach, "Experts and Trump's Advisers Doubt White House's 240,000 Coronavirus Deaths Estimate," *Washington Post*, April 2, 2020. William Wan and Carolyn Y. Johnson, "America's Most Influential Coronavirus Model Just Revised Its Estimates Downward. But Not Every Model Agrees," *Washington Post*, April 6, 2020.

230 **seven states still had no statewide stay-at-home orders**: Sarah Marvosh, Denise Lu, and Vanessa Swales, "See Which States and Cities Have Told Residents to Stay at Home," *New York Times*, April 20, 2020.

230 **By late April, several states had begun**: Associated Press, "U.S. States Represent Patchwork as They Mull Economic Restarts," *Washington Post*, April 22, 2020.

230 **The model got the desired response**: Sheri Fink, "White House Takes New Line After Dire Report on Death Toll," *New York Times*, March 16, 2020, updated March 17, 2020.

230 **put out new, more optimistic projections**: Wan, Dawsey, Parker, and Achenbach, "Experts and Trump's Advisers Doubt."

231 **IHME modelers revised their death estimates:** Philip Bump, "The Lesson of Revised Death Toll Estimates Shouldn't Be That Distancing Was an Overreaction," *Washington Post*, April 9, 2020.

231 **People of different political persuasions**: Quotations of Fauci and Kilmeade in Bump, "The Lesson of Revised Death Toll Estimates."

232 "**WE CANNOT LET THE CURE BE WORSE**": Trump's tweet quoted in Maggie Haberman and David E. Sanger, "Trump Says Coronavirus Cure Cannot 'Be Worse Than the Problem Itself,'" *New York Times*, March 23, 2020.

232 **"Why is nobody putting some numbers on"**: Walter Scheidel, economic historian at Stanford University, quoted in Porter and Tankersley, "Shutdown Spotlights Economic Cost of Saving Lives."

233 **Several European governments**: Peter S. Goodman, "The Nordic Way to Economic Rescue," *New York Times*, March 28, 2020; Emmanuel Saez and Gabriel Zucman, "Jobs Aren't Being Destroyed This Fast Elsewhere. Why Is That?" *New York Times*, March 30, 2020.

233 **When they apply cost-benefit analysis**: Cass R. Sunstein, *Valuing Life: Humanizing the Regulatory State* (Chicago: University of Chicago Press, 2014).

233 **The field of cost-benefit analysis has a long tradition**: Frank Ackerman and Lisa Heinzerling, *Priceless: On Knowing the Price of Everything and the Value of Nothing* (New York: New Press, 2004), ch. 3; and Sunstein, *Valuing Life*, 86–87, 102–3, and 116.

233 **the Council of Economic Advisors**: Jim Tankersley, "White House Economists Warned in 2019 a Pandemic Could Devastate America," *New York Times*, March 31, 2020.

234 **"The fundamental problem is"**: Tweet by Scott McMillan, quoted in *Washington Post Daily 202*, March 26, 2020.

234 **The lieutenant governor of Texas**: Dan Patrick made this suggestion on Tucker Carlson's Fox News show, March 24, 2020.

234 **"My mother is not expendable"**: Cuomo's press briefing, March 24, 2020.

235 **Dan Wikler, a medical ethicist**: Dan Wikler, "Here Are Rules Doctors Can Follow When They Decide Who Gets Care and Who Dies," *Washington Post*, April 4, 2020.

237 **"Why don't we know"**: Ibram X. Kendi, "Why Don't We Know Who the Coronavirus Victims Are?" *Atlantic*, April 1, 2020.

237 **The CDC wasn't reporting**: Spencer Overton, "CDC Must End Its Silence on the Racial Impact of COVID-19," *Washington Post*, April 7, 2020.

237 **When a team of reporters**: Kat Stafford, Meghan Hoyer, and Aaron Morrison, "Outcry Over Racial Data Grows as Virus Slams Black Americans," *Washington Post*, April 8, 2020.

237 **Blacks, Hispanics, and in some places Asians**: Ibram X. Kendi, "What the Racial Data Show," *Atlantic*, April 6, 2020; Meaghan Flynn, "'Those Numbers Take Your Breath Away': COVID-19 Is Hitting Chicago's Black Neighborhoods Much Harder than Others, Officials Say," *Washington Post*, April 7, 2020; Stafford, Hoyer, and Morrison, "Outcry Over Racial Data."

238 **"Without racial data"**: Kendi, "Why Don't We Know."

239 **Whites had tied some of them**: Richard Rothstein, *The Color of Law* (New York: Liveright, 2017).

239 **opportunity to expand Medicaid under Obamacare**: Jamila Taylor, "Racism, Inequality, and Health Care for African Americans," The Century Foundation, December 19, 2019.

240 **Mayor Lori Lightfoot**: Flynn, "'Those Numbers Take Your Breath Away.'"

241 **"'I learned,' he says, 'that I had had'"**: Albert Camus, *The Plague* (New York: Vintage International, 1991), pp. 251, 253.

241 **"They are numbers, and they are facts"**: Cuomo's press briefing, April 29, 2020.

Illustration Credits

x © Dave Coverly / speedbump.com.

6 (left) © 2013, 2017, 2019 Sesame Workshop®. Sesame Street® and associated characters, trademarks, and design elements are owned and licensed by Sesame Workshop. All rights reserved.

6 (right) © 2013, 2017, 2019 Sesame Workshop®. Sesame Street® and associated characters, trademarks, and design elements are owned and licensed by Sesame Workshop. All rights reserved.

7 Charles E. Siler aka Chuck Siler.

25 Leo Cullum, The Cartoon Bank © Condé Nast.

39 Bernard Schoenbaum, The Cartoon Bank © Condé Nast.

54 (top) Steve Kelling, the Cornell Lab of Ornithology.

54 (bottom) Steve Kelling, the Cornell Lab of Ornithology.

56 © Dave Coverly / speedbump.com.

68 Barbara Smaller, The Cartoon Bank © Condé Nast.

90 Dana Fradon, The Cartoon Bank © Condé Nast.

100 *The Boston Globe.*

124 © Joel Pett. Reprinted with permission.

141 Emily Flake, The Cartoon Bank © Condé Nast.

152 © Joel Pett. Reprinted with permission.

176 Data visualization created for the National Registry of Exonerations by Dustin Cabral. © 2020 National Registry of Exonerations, updated to March 18, 2020.

182 © Dave Coverly / speedbump.com.

199 Shuhada Organization, Afghanistan.

202 Illustration from *Would You Kill the Fat Man? The Trolley Problem and What Your Answer Tells Us about Right and Wrong* by David Edmonds. Copyright © 2014 by David Edmonds. Reprinted by permission of Princeton University Press.

Index

Page numbers in *italics* refer to figures. Page numbers followed by n refer to notes.